THE SAGE GROUSE
IN WYOMING

The
SAGE GROUSE
in Wyoming

By ROBERT L. PATTERSON

Sketches by Charles W. Schwartz

Reprint of 1952 Edition by Wyoming Game and Fish Commission

The Sage Grouse in Wyoming

ISBN-10: 1-932846-31-X
ISBN-13: 978-1-932846-31-7

Library of Congress Control Number: 2017963165

THE BLACKBURN PRESS
P. O. Box 287
Caldwell, New Jersey   07006   U.S.A.
+1 973-228-7077
www.BlackburnPress.com

One of my most vivid recollections of early childhood is of standing as a partly clad toddler in the open doorway of my parents' mountain homestead cabin near the western boundary of Wyoming, pleasantly warmed by the thin spring sunshine and gazing out upon a flock of sage grouse trailing down a sagebrush hill beside the house, heading for an open field beyond. It was strutting season, and the picture of the pompous cocks parading in military fashion with chests blown up and tails fanned out was etched so deeply on my memory that time has failed to blur it.

Another impression that continues clear is one of racing my pony across a wide clay flat where hundreds of "sage chickens" had congregated during mating season, shouting and whistling to "whoop them up" and see them scatter.

Some years later, I remember driving a wagon behind an advancing line of hunters as they combed a field in summer, the wagon being required to haul the "bag." I remember helping to skin and dismember the big, turkey-like carcasses, saving only the breasts and legs for pickling in brine to prevent spoilage in hot weather.

Still later, as a deputy game warden, I remember when the advent of automobiles and good roads spread sage grouse hunting from the easily accessible areas to the wildest and most remote sagelands—wherever the bird was found. I saw its native range disrupted by encroaching civilization, and saw it hunted in midsummer seasons when the young could not fend for themselves. From populations so abundant as to be considered inexhaustible, I saw their numbers decline to a point where it was necessary for the game and fish commission of Wyoming and other western states to prohibit hunting altogether. For several years after complete protection was thus afforded, I heard well-informed wildlife experts profess that protection had come too late and gloomily predict that the sage grouse, like the buffalo, had outlived its era and was doomed to disappear as a game species.

In 1937 Wyoming followed the example of other states and temporarily banned sage grouse shooting within its borders. Considerable concern was felt when the birds did not respond visibly to removal of hunting pressure, and various theories were advanced in explanation of their failure to recover. Disease and predation were strongly suspected; and in December of 1939 the Wyoming Game and Fish Commission inaugurated a preliminary study of the grouse, in co-operation with the Federal Aid in Wildlife Restoration program, in part to investigate the effects of these factors on vitality of the species. Another major purpose of the project was to restore the bird by live-trapping and transplanting to habitat where it had become depleted.

Wartime dislocations interrupted the study before any conclusions had been reached, and before any appreciable dispersion of the species had been accomplished. In 1948, after various personnel changes had hampered operations, Robert L. Patterson was employed by the Wyoming Game and Fish Commission to reorganize and direct the study which was completed early in 1952.

To my knowledge, this is the first definitive study ever made of the most magnificent of the several species of western grouse, and the material contained and conclusions reached have such timely significance that the Game and Fish Commission deemed it worthy of wide circulation. Again, through the co-operation of the Federal Aid in Wildlife Restoration program, it has been possible to publish the study findings in the hope that other western states with similar sage grouse problems can benefit from our own experiences in attempting to restore this splendid native bird.

It is also hoped that the book will have interest and significance to persons other than scholars and wildlife management officials, for it deals with a subject that is of consummate interest to every person in the West. The great sage grouse is the most important of the upland game birds indigenous in this sector of the continent, and is as distinctively "western" as the Stetson hat and the sagebrush. It once was as abundant and as widely dispersed as the sagebrush itself and was so spectacular in its behavior that it invariably captured the imagination of early travelers and residents of the intermountain plains and valleys. All western Indian tribes had dances that depicted its movements during the mating season; western explorers and immigrants were almost as intrigued by it as by the comparably numerous buffalo; and early journals and diaries devoted much space to both. The buffalo's contribution to the larders of both Indians and the first white settlers is well-understood, but that made by the sage grouse has never been fully recognized.

Ever since its initial discovery by Lewis and Clark in 1806, the sage grouse has stimulated the interest and whetted the curiosity of all observers. Perhaps no other species, through the years, has been subjected to closer and more protracted attention. Yet, paradoxically, it has remained essentially a bird of mystery, so novel and unusual in many of its behavior patterns that—intensive observation notwithstanding—it still is the object of more speculation and improbable folklore than any other well-known wild creature.

This comprehensive treatise includes the first complete study of the natural history of the bird, and no doubt will do much to help dispel the cloud of superstition and false belief which in the past has been fostered even by supposedly qualified naturalists, thereby complicating still further the task of intelligent management.

In addition, the investigation findings have pointed up the actual factors governing sage grouse vitality and population trends. With this information in hand, I now am confident that the dour prophecies of a decade ago regarding the future of the bird will not be fulfilled. Already, it is maintaining itself on some portions of its range so successfully that a resumption of open hunting seasons had to be ordered in recent years in order to prevent overpopulation, and subsequent damage to the bird as well as its environment.

As recently as 1936, Dr. Aldo Leopold, eminent American wildlife authority, was quoted as saying: "No comprehensive life-history study of a distinctively western game bird has ever been made. Such a study would have a future influence comparable to that of the Georgia Quail Investigation in the East. It might leaven the future of wildlife conservation in the whole western region."

It is modestly but confidently hoped that, with the publication of this distinguished monograph, the omission cited by the late Dr. Leopold has found remedy.

LESTER BAGLEY
Wyoming Game and Fish Commissioner

PREFACE

The Sage Grouse in Wyoming is the product of research sponsored by the Wyoming Game and Fish Commission. This work was initiated in 1940 with Federal Aid to Wildlife Restoration funds under Pittman-Robertson Project Number 28-R.

The author has been associated with this undertaking continuously as project leader from May, 1948 until its completion in March, 1952. The scope of the studies widened considerably in 1948 with a view to preparing a monograph on the sage grouse. Organization and development of the field studies as well as the preparation of the findings for publication have been solely the responsibility of the writer since his employment with the department in 1948. This publication is the result of individual effort, although the information contained in quarterly progress reports of previous years was utilized, wherever possible, to supplement the data collected during the past four years.

The present study has been concerned primarily with year-round sage grouse investigations conducted on an intimately known area of land. Part I is devoted to a discussion of the environmental factors affecting sage grouse abundance, including a detailed account of the restoration of the species in recent years. Natural history aspects and general behavior of grouse populations were thoroughly examined, and these findings appear in Part II. The effects of man's various land-use activities on sage grouse numbers merited comprehensive study, and Part III is comprised mainly of an evaluation of these factors. This treatise on the sage grouse concludes with a summary of management recommendations, including a forecast for the bird's future.

It is anticipated that some criticism will be directed against the elements of this study which depart from purely natural history and ecological phases. The most exacting knowledge of an animal's life processes by itself is valueless in an evaluation of its chances for survival in the atomic age. The various land-use policies and political expediencies ultimately set the tolerance limits for the survival of any wildlife species. The sage grouse provides no exception to this principle. This monograph would have been erroneously conceived and derelict in its presentation if the effects of an expanding civilization upon sage grouse populations had not been fully explored and appraised.

The author wishes to acknowledge with utmost respect and gratitude the unfailing support and enduring patience exhibited by the individual members of the Commission during the progress of the study. State Game and Fish Commissioner Lester Bagley gave unlimited encouragement to the preparation of this monograph.

Warren J. Allred and Floyd M. Blunt supervised the various aspects of the investigation as co-ordinators of the Pittman-Robertson wildlife restoration program in Wyoming, and gave unstintingly of the materials and resources at their command.

The field force of the game division merits particular appreciation for their loyal assistance in gathering data at game checking stations and for making available information on the distribution of sage grouse in Wyoming. District Supervisor C. W. Sheffner and Deputy Game Warden Dewey Henderson were especially helpful. Assistance in the execution of the trapping and transplanting program as well as in the development of trapping techniques was received from field assistants Euvern F. Putnam and H. B. Sanderson.

The co-operation of the office force of the department has been absolute and unreserved. Chief Clerk S. J. Jiacoletti and staff actively assisted in the preparation of the final manuscript, and individually they deserve much recognition.

For special help on technical problems associated with the study, the author is indebted to the following individuals and governmental agencies:

For identification of food habits material: Leon H. Kelso, H. W. Norton, N. Hotchkiss, and R. T. Mitchell of the Food Habits Section of the U. S. Fish and Wildlife Service.

For examination of diseased material and for quantitative analyses of food materials in sagebrush leaves and grouse droppings: Ralph Honess, Kenneth Winter, and George Post of the Wyoming Game and Fish Commission Disease Research Laboratory at Laramie.

For furnishing data on mammal ranges and nomenclature: Rollin Baker, University of Kansas; and Arthur B. Mickey, University of Wyoming.

For information on reclamation developments in the Green River Basin, for the use of unpublished data on geological and soil conditions in Eden Valley, and for assistance in the preparation of base maps: the U. S. Bureau of Reclamation and the U. S. Soil Conservation Service.

For identification of plants: C. L. Porter, Curator, Rocky Mountain Herbarium.

For carefully reviewing the manuscript and for offering suggestions for its improvement: Lester Bagley, Wyoming Game and Fish Commission; Dee Linford, western free-lance writer; various personnel of the U. S. Fish and Wildlife Service; and Loula Mai Patterson.

For technical assistance in the development of the project from 1948 until 1952 and for considered advice in the preparation and editing of the final manuscript: Warren W. Chase, Elzada U. Clover, Samuel T. Dana, Samuel A. Graham,

William H. Burt, and Karl F. Lagler of the graduate faculty of the University of Michigan.

The suggestions of James R. Simon and Richard Pough were helpful in conducting phases of the field studies.

Both as a layman and a livestock man, John A. Coppes, Farson, Wyoming, followed the evolution of the study with keen interest and contributed invaluable help in his many excursions with the writer into unmapped areas of the Green River Basin. Hazards involved in winter reconnaissance of mountainous and desert regions of the Basin were materially reduced through his thoughtfulness and generosity.

Lastly, the author wishes to express his sincere appreciation to the personnel of the game and fish departments of the western states and provinces who are deserving of no small amount of gratitude for their substantial help in contributing material for this dissertation.

Sage grouse study skins prepared on this project have been deposited in the Department of Zoology of the University of Wyoming at Laramie.

Unless otherwise credited, all photographs were taken by the author.

March 6, 1952   ROBERT L. PATTERSON

TABLE OF CONTENTS

LIST OF TABLES

TABLE OF FIGURES

TABLE OF ILLUSTRATIONS

4. Clutches with five eggs usually indicate a renesting attempt
5. Grouse nests are frequently placed under a sagebrush limb
6. Destruction by ground squirrels accounted for more than 40 percent of all nests found destroyed
7. Badgers caused a 36 per cent nesting loss
8. Rabbitbrush is occasionally utilized as nesting cover
9. Pintail duck nest situated on a sage grouse nesting plot more than one mile from water
10. Sage grouse nest and eggs after a May snowstorm
11. No nesting mortality occurred during the study from cold weather and snowstorms
12. An occasional hen displays no fear of humans
13. Nest desertion was improbable during the late stages of incubation
14. Ninety-two per cent of all nests observed were located under sagebrush
15. Nests are frequently well-concealed in dense sage-brush clumps
16. Incubating hens are well-camouflaged regardless of nest site
17. Sage grouse incubation period varied from 25 to 27 days in Eden Valley
18. Average nesting densities varied from 42 to 59 nests per square mile
19. The nature of sage grouse courtship and breeding behavior prevents repeated nesting attempts
20. A ruffled sage hen defends her eggs and newly hatched chick
21. Newly hatched grouse chicks and nest site
22. Large "clocker" or incubation dropping adacjent to a nest of hatched eggs
23. Ten-day-old sage grouse chicks showing aluminum wing clip used in marking young birds
24. Day-old sage grouse chicks

1. Cocks commence strutting activities in January on winter roosting sites
2. A few cocks move onto snow-covered strutting ground in late February
3. The immature cock may be recognized by the narrow white band on the cape
4. The size of strutting ground plots varies
5. During a phase of their territorial defense a pair of cocks may engage in extended clucking duels prior to actual combat
6. Cocks defend individual strutting territories by bluffing displays and actual combat
7. The dominant cock is frequently surrounded by several "guard cocks"
8. The majority of matings take place during morning twilight
9. Hens appear on strutting ground in large numbers in early April
10. In mid-April hens concentrate on strutting ground mating spots
11. A dominant cock actually performs the majority of mating acts on a given mating spot
12. Hens signify their readiness for mating by squatting in front of a cock
13. A satiated cock may stand on a hen's back for several seconds after the mating act
14. Peak of mating activity occurs during the third week of April in Eden Valley
15. Swishing sounds are produced by wings rubbing against the stiff, white feathers on the cape
16. Grouse in Eden Valley terminate seasonal strutting activities in late May
17. Sage grouse cocks on winter range near Eden Valley
18. Spring snowstorms seldom disrupt strutting activities
19. Cocks utilize the same strutting ground throughout the breeding season

20. An imature male cautiously approaches the strutting domain of an old cock
21. A 7 lb. adult cock trapped on strutting ground
22. Yearling cock and adult cock trapped on strutting ground
23. Development of the breeding plumage in the immature cock
24. The mature breeding cock is twice as large as the hen

GROUP IV: TRAPPING AND TRANSPLANTING   *following page 236*

1. Sage grouse traps are constructed of 2-inch mesh, cotton netting erected on a rectangular frame
2. The net structure is supported by three transverse arches
3. Posts are 52 inches long and pointed
4. Trap is secured to ground at 3-foot intervals
5. For strength and stability the netting is faced along the edges with 1/4-inch rope
6. Each trap is equipped with two wings of 2-inch cotton netting
7. Entrance panels are attached to the front corners of the trap
8. Grouse are transplanted to release sites in bird farm crates

GROUP V: AGE DETERMINATION AND DIGESTIVE SYSTEM
following page 252

1. A juvenile hen and an adult hen in October
2. A juvenile hen and a juvenile cock in October
3. A mature hen in October
4. A juvenile hen in October
5. A juvenile cock in October
6. Digestive system of the sage grouse
7. Cross section of a sage grouse gizzard
8. Crop and gizzard of a sage grouse

1. Sage grouse breeding ranges in the mountain foothills are covered with deep snow in the winter
2. In the winter sage grouse are found in regions of lighter snowfall and exposed sagebrush on the semi-arid deserts
3. The vast majority of Wyoming's antelope herds are located on sagebrush-grass range types
4. Antelope share seasonal ranges with sage grouse in Wyoming
5. Sagebrush is probably the most important winter food of mule deer in the intermountain region
6. Cottontail rabbits rely heavily upon sagebrush types for their habitat requirements in many parts of the West
7. Several hundred thousand sheep annually trail across sage lands in the Green River Basin
8. Domestic sheep in the intermountain region depend upon sagebrush-grass types for their forage requirements from five to seven months out of the year
9. Sustained production of livestock as well as wildlife is vitally dependent upon preservation and careful management of sagebrush lands in the Mountain West
10. Grazing of livestock constitutes an important use of sagebrush lands in Wyoming
11. World's largest oil rig and deepest well on Dry Sandy—Pacific Creek sage grouse study tract
12. Airstrip on Dry Sandy—Pacific Creek area
13. Elimination of sagebrush habitat poses a constant threat to sage grouse survival in many parts of the birds' range
14. Demonstration area mowed in July, 1950, and drilled to crested wheatgrass in fall of 1950
15. Improvised mowing machine used in sagebrush eradication and control programs
16. The McCoy sweep designed to cut sagebrush roots

THE SAGE GROUSE
AND ITS ENVIRONMENT

THE SAGE GROUSE IN THE WEST

Introduction

Members of the Lewis and Clark expedition have generally been credited with the first discovery of the sage grouse by white men. Bonaparte (1827) named this new species, *Tetrao urophasianus,* and supplied the first technical description, an account of which follows:

Repeated accounts of hunters and travelers have long since put beyond question the existence, in the western wilds of the United States territory, of a very large species of Grouse, analogous to the European *Tetrao urogallus* (capercaillie); which, however, not having fallen under the immediate inspection of any naturalist, could not be properly registered in systematic catalogues. Having had the good fortune to find a specimen of this most interesting species among the endless ornithological treasures, of which Mr. Leadbeater is so liberal towards the lovers of science, I beg leave to introduce it in its proper place, as one of the chief ornaments of the North American Fauna.

Genus. *TETRAO*, L.

Sub-genus, *TETRAO*, Nob.

Tetrao urophasianus

Head smooth; primaries unspotted; toes strongly pectinated; tail subcuneiform, of twenty narrow tapering feathers.

Male black? Female grey, mottled.

Inhabits the North Western countries beyond the Mississippi, especially on the Missouri.

Size of the *T. urogallus*, which species it represents in the New Continent.

Aboriginal man had doubtlessly been familiar with this grouse for unknown ages prior to its discovery by white men. Indian tribes were so impressed with the bird's abundance in parts of western Wyoming that they named a major river system in its honor. As late as 1840 the Green River was commonly called the Seeds-ke-dee-agie, or Prairie Hen River, obviously referring to the sage grouse.

In terms of geologic time it is reasonable to assume that considerable areas of western United States and Canada have been covered with shrubs of the climax sagebrush type for countless eras. The present range of the sage hen is limited by the distribution of the climax sagebrush type. The only definite information relative to the age of the sage hen as a species is obtained from forty-four bones recovered from Fossil Lake, Oregon. This material is of Pleistocene origin and, according to Wetmore (1951), does not differ in any appreciable way from the skeleton of the modern bird.

Taxonomy

Classification

The American Ornithologists' Union has assigned the name of sage hen to the species. It is more commonly referred to as the sage chicken by the local populace. A considerable number of naturalists and wildlife biologists refer to the bird as the sage grouse. Within the text of this treatise I shall apply all of these accepted common names synonymously to the bird.

In 1831 Swainson framed for this bird the genus *Centrocercus*, and thus the presently accepted scientific name of *Centrocercus urophasianus,* or as derived from the Latin, a spiny-tailed pheasant.

The sage hen belongs to the order of gallinaceous birds which is characterized by the presence of a specialized stomach, or gizzard. This group of fowl-like birds includes the turkey, partridge, quail, pheasant, grouse, jungle fowl, and domestic chicken. All of these birds possess digestive systems which are adapted to a diet of hard seeds, herbs, and browse. The order, *Galliformes,* is represented by several families of gallinaceous birds. The family *Tetraonidae* embraces the grouse-like birds such as the sage hen, ruffed grouse, prairie chicken, and sharp-tailed grouse. In common with other members of the family the sage hen possesses a completely feathered head, except for a bare strip of skin above the eye; an external nasal opening which is completely feathered; a tarsus which is feathered to the toes on the upper side; and no spurs. The sexes are similar and have bare spots on the front of the neck and long pointed tails.

Description

The sage hen is a large grayish-brown grouse of the sagebrush prairie, as large as a domestic chicken, identified by the prominently dark belly and long tapering tail feathers. The males are nearly twice as large as the females.

The following detailed description was furnished by Bonaparte:

> Adult male: The bill of the adult male is short and black, and the area immediately bordering the lores is densely feathered with short, buffy, gray-tipped, black feathers. The iris of the eye is brown, and the pupil a shiny black. Above each eye is a small orange-colored superciliary membrane or tissue. The crown is a mixture of black, gray, and tan; a white superciliary line extends from in front of the eye back to the ear opening. The ear coverts are gray and brown filoplume feathers. The backs of the head and nape are gray, with an occasional line of black or tan. The black and white blotched chin is separated from the throat by a narrow white collar, extending from the eye down under the head. The upper throat is covered with gray and white-tipped, black feathers, which increase in size as they extend down the neck. The sides of the neck are gray and white-trimmed, with long filoplume feathers which are white at the base; the distal half is very narrow and black, with an

occasional small white tip. On the fore-neck are some loose rolls of skin, which function as air sacs or pouches, covered with short, stiff, scale-like, white feathers. These specialized feathers surround two olive-colored bare spots, one on each side of the throat. The breast feathers are white with a black rachis and a black tip. As these feathers extend down the breast toward the abdomen, they become larger, and the amount of black on the tip increases, until when the abdomen is reached, they appear to be solid black. The abdomen is black in appearance; however, the basal portion of these feathers is white or gray. The back sides are black, marked with bars and spots of white, gray, tan, and brown. The rachis and central line of the side and flank feathers are white. The wings are grayish-brown, with a slight mottling of tannish-gray. The primaries are solid grayish-brown. The under surface of the wings are white. The legs and tarsi are densely covered with soft, downlike, gray feathers, which extend down to the toes. The toes are black, and on each side are small narrow scales, which act as snow webs during the winter. The upper tail feathers are black and gray, blotched with light shades of brown. The tail feathers are pointed in the nuptial plumage, as well as at all other times. The number of tail feathers in the males that appeared to have a complete plumage, varied from 18 to 21. The under-tail coverts are glossy black, tipped with pure white round spots. The tail, when in natural position, tapers to the distal end.

Adult female: The adult female is similar to the male, but is smaller and more drab in appearance. The abdomen is not as solid black in appearance as is that of the male. The abdomen feathers are at times edged with white. The breast is white, with blotches of tan, gray, and black. The female does not have air pouches nor the specialized white feathers on the fore-neck. The filoplume feathers on the sides of the neck are also lacking. The back, rump, flanks, and tail are lighter in color, with more gray and tan markings and less black. The tail feathers are much shorter and somewhat lighter than are those of the males.

Juveniles: The natal plumage, or down, is a mottled grayish-black. The juvenile is similar to the adult females; however, the breast feathers of the juveniles are lighter in color, there being more brown, and less black. The entire plumage of the juveniles has more tan, gray, and brown, and less white and black than has that of the adult female.

Sexual Differences

With the exception of the sage grouse, external differences in sex are not conspicuous among North American grouse species. During most seasons of the year careful examination is required to distinguish the sexes among ptarmigan, ruffed grouse, dusky grouse, prairie chickens, sharp-tailed grouse and others. It is difficult to distinguish the sex of grouse while in flight.

The sage hen possesses distinctive external characters for separating the sexes. Sex of adult sage grouse may be readily determined by skilled observers at all seasons. Both sexes possess characteristic features of coloration, gross appearance, and action in flight which facilitate identification. The sex of juvenile sage grouse is not readily determined until the adult plumage commences to appear late in the fall.

During the course of this project, more than 6,000 adult and juvenile sage grouse were handled in connection with live-trapping and transplanting operations. It has been the author's good fortune to handle over 3,000 birds during his tenure with this study from 1948 to 1951. Live-trapping and banding operations afforded excellent opportunities for observing sex and age differences, and for studying growth, plumage development, and behavior patterns.

As an aid in portraying the external differences in sex and age, the author has prepared a series of sage grouse study skins, photographs of which are included in this report.

In the breeding plumage the sexes may be distinguished externally by the use of one significant feature: the appearance of the throat and neck. On the male the predominantly dark chin and throat is separated from the black upper neck by a narrow white collar. The lower neck is covered on the front and sides by short, stiff, scale-like, white feathers which gradually merge with the breast feathers. The latter are not pure white but possess a black rachis and black tip. The sides of the neck are adorned with several filoplume feathers, 4 to 6 inches long, which are white at the base and tipped with black. The general color pattern on the male from chin to abdomen is one of contrasting zones of black and white feathers.

The female, on the other hand, features a grayish-white chin and upper throat, in marked contrast to the male's black chin and throat. The entire neck and breast region is covered with a solid pattern of feathers blotched with tan, gray, and black. The female lacks the specialized stiff white feathers and filoplumes on the neck.

In the non-breeding plumage, the male acquires a grayish-white chin and upper throat similar to that of the female. The filoplumes are lacking. The white collar is retained. The black band on the upper throat is less prominent but, nevertheless, conspicuous. The specialized stiff white feathers on the neck are partially obscured by softer feathers but are evident upon close examination. In varying degrees, the black and white zonation pattern is characteristic of the adult male in all plumages.

The plumage pattern of the adult female exhibits little variation through the year.

Another diagnostic feature of the bird's plumage is the variation in the length of the tail. The mature male has a tail which generally exceeds 8¾ inches, whereas the tail of a mature hen seldom exceeds 8¼ inches. On the basis of tail lengths and plumage, it is possible to distinguish adult and immature cocks during the winter, spring, and early summer months. Excluding males in their first winter plumage, the average tail length of an adult cock is 11½ inches, in contrast to an average tail length of 7½ inches for an adult hen. The immature male does not acquire full adult plumage until its second winter of life. Due to longer length, the tail of an adult cock drags against the ground as the bird alights, causing the feathers to become worn and prominently pointed at the tip.

The olive-colored bare spots over the air sac region of males are visible in all plumages. Similar spots are present in females, but are much smaller and concealed in the neck feathers.

Gross Appearance

During the winter and spring seasons, the adult male is quickly recognized by the striking white nuptial plumage on the neck and breast regions. Furthermore, at these periods the adult cock will average 3 pounds heavier than the drably-colored hen, and appear twice as large as she. The average breeding season weight for an adult rooster is about 6 pounds; a hen will only average half as much.

The heavier-bodied male walks with a slight waddling motion, whereas the female typically exhibits a steady, even stride.

Specialized activity is associated with the male during the breeding season. Except for isolated occurrences involving an occasional female, only the male bird engages in strutting and courtship displays. Cocks neither assume a role in nesting activities, nor in rearing the young.

The sex of sage grouse can usually be determined on the basis of flight patterns. The hen rises quickly from the ground and alternately dips her body from side to side in the course of flight. The male rises laboriously, like a heavy bomber, and maintains an even flight once leveled off. In flight, both sexes alternately flap their wings rapidly for several beats, and then glide for considerable distances.

The sexual differences of sage grouse are summarized in Table 1.

Range

On the basis of habitat requirements, three species of grouse are considered as "prairie grouse." Sharptails inhabit the semi-prairies dominated by grasses, shrubs, and open woodlands. Prairie chickens characteristically are found on the true grass prairies where there are scattered groves of trees. Sage grouse are representative of the treeless, sagebrush-grass prairies.

The original range of the sage grouse closely conformed to the distribution of the sagebrush, *Artemisia tridentata* Nutt. and related species. This area included the semiarid plains of the intermountain and northwestern states and the southern border of the three southwestern Canadian provinces. The sagebrush prairie has developed as a climax vegetative type and may be classified as a true primeval wilderness. It originally extended from western Kansas, western Nebraska, and western Dakotas, westward to include northeastern Arizona, northwestern New Mexico, the northwestern sections of Colorado, and practically all of Montana, Idaho, Wyoming, Nevada, and Utah, excluding

TABLE 1
Comparison of Sex Differences in Sage Grouse

	Male	Female
Appearance	Averaging twice as large and heavy; striking black and white plumage on neck and breast regions.	Noticeably smaller; solid color pattern on neck and breast.
Head	Averaging 3¼ in. from back of skull to tip of bill.	Averaging 2¾ in. from back of skull to tip of bill.
Tail	More than 8¾ in. long; feathers worn at tips, exposing rachis.	Less than 8¼ in. long; feathers tapering normally and seldom worn at tips.
Toes	Middle toe longer; averaging more than 2½ in. in length.	Middle toe shorter; averaging less than 2¼ in. long.
Leg	Tarsus averaging more than 2½ in. long.	Tarsus averaging less than 2¼ in. long.
Walk	Walks with a slight waddling gait.	Walks with an even, steady stride.
Flight	Takes off laboriously and maintains a horizontal equilibrium.	Takes off quickly and dips body alternately from side to side.
Weight	Varies from 5 lb. to 7 lb. in winter and spring; averaging slightly more than 4½ lb. at other seasons.	Varies from 2 lb. 14 oz. to 3 lb. 6 oz. in winter and spring; averaging slightly more than 2½ lb. at other seasons.
Breeding	Distinctive courtship displays; does not assist in nesting or rearing of the young.	Passive role in courtship; performs all of the nesting duties.

the forested mountainous regions. It also extended into the east-central portion of California, the eastern halves of Oregon and Washington, and the southern border of the three southwestern Canadian provinces. Within this extensive area two distinct range types (the sagebrush-grass and salt-desert shrub) furnished the habitat requirements of the sage grouse.

In connection with the forthcoming revision of the A. O. U. Check List, the range outline for sage grouse as given in the 4th edition has been modified as follows:

> *Resident,* locally, now in reduced number, from central Washington, eastern Oregon, southern Idaho, eastern Montana, southeastern Alberta, southern Saskatchewan, and western North Dakota (Billings County), south to central-eastern California, south-central Nevada, Utah, western Colorado, and northwestern Nebraska (Sioux and Dawes counties); formerly to southern British Columbia (Osoyoos Lake), northern New Mexico (Tres Piedras and Tierra Amarilla), and the Oklahoma panhandle (Wetmore, op. cit.).

The sagebrush-grass range now covers more than 90 million acres. It is typical of rather dry valleys and intermountain basins where most of the meager annual precipitation occurs during spring and winter months, varying from 5 inches on the prairies to 30 inches in the foothills. Elevations in this type range chiefly between 2,000 feet (central Washington) and 8,000 feet (Wyoming). Most of the cultivated area in this region was formerly sagebrush-grass type.

The salt-desert shrub range presently extends over 40 million acres in the arid parts of Nevada, Utah, eastern Oregon, western Colorado, and southwestern Wyoming. The precipitation varies from 5 to 10 inches yearly. The soils are saline in character and the plant cover is sparse.

Distribution and Relative Abundance

In primitive times, sage grouse were found throughout most of the range and habitat previously described. However, it is doubtful if sage grouse have occurred in recent times in Kansas or Arizona. Sage grouse distribution was discontinuous, due to the interspersion of forested mountainous areas, although, as will be shown later, mountains were no barrier to the dispersal or movement of sage chickens within their natural range. Historically, as today, sage grouse were most commonly encountered near the mountains. High plateaus and intermountain valleys provided the best conditions for the continued survival of sage grouse populations.

It has been estimated by Martin, Zim, and Nelson (1951) that more than 50 per cent of the original sage grouse habitat has been eliminated. In spite of the great reduction in habitat which resulted from the settlement and development of western United States and Canada, the geographical distribution of the species has remained relatively unchanged. Perhaps only in British Columbia has the species been totally exterminated. It was inevitable, however, that a reduction in numbers of sage grouse should accompany this major change in the pattern of land use. In an effort to determine the present status of sage grouse throughout its range, the aid of the state and provincial game departments has been solicited. As a result of their cooperation, it has been possible to prepare a composite map showing the present distribution and relative abundance of sage grouse throughout the West (Figure 1).

Status of Sage Grouse as a Game Bird

Within the space of comparatively few years the sage grouse descended from a position of leadership as a game bird in several western states to a point where its preservation from extinction was considered doubtful by many conservationists and wildlife biologists. In 1937, Girard implied that the impending fate of sage grouse was extinction and stressed the need for immediate action to save the species.

In pioneer times this bird was considered the leading upland game bird in nine western states. With the settlement of the West and the inevitable intensification of agricultural, grazing, and hunting activities, the sage grouse diminished to a point where in the five years prior to 1937 it was considered an important game bird only in parts of Montana, Wyoming, Idaho, and Nevada. In 1937 the hunting of sage grouse was discontinued in Wyoming and greatly restricted seasons were instituted in Nevada (Rasmussen and Griner, 1938).

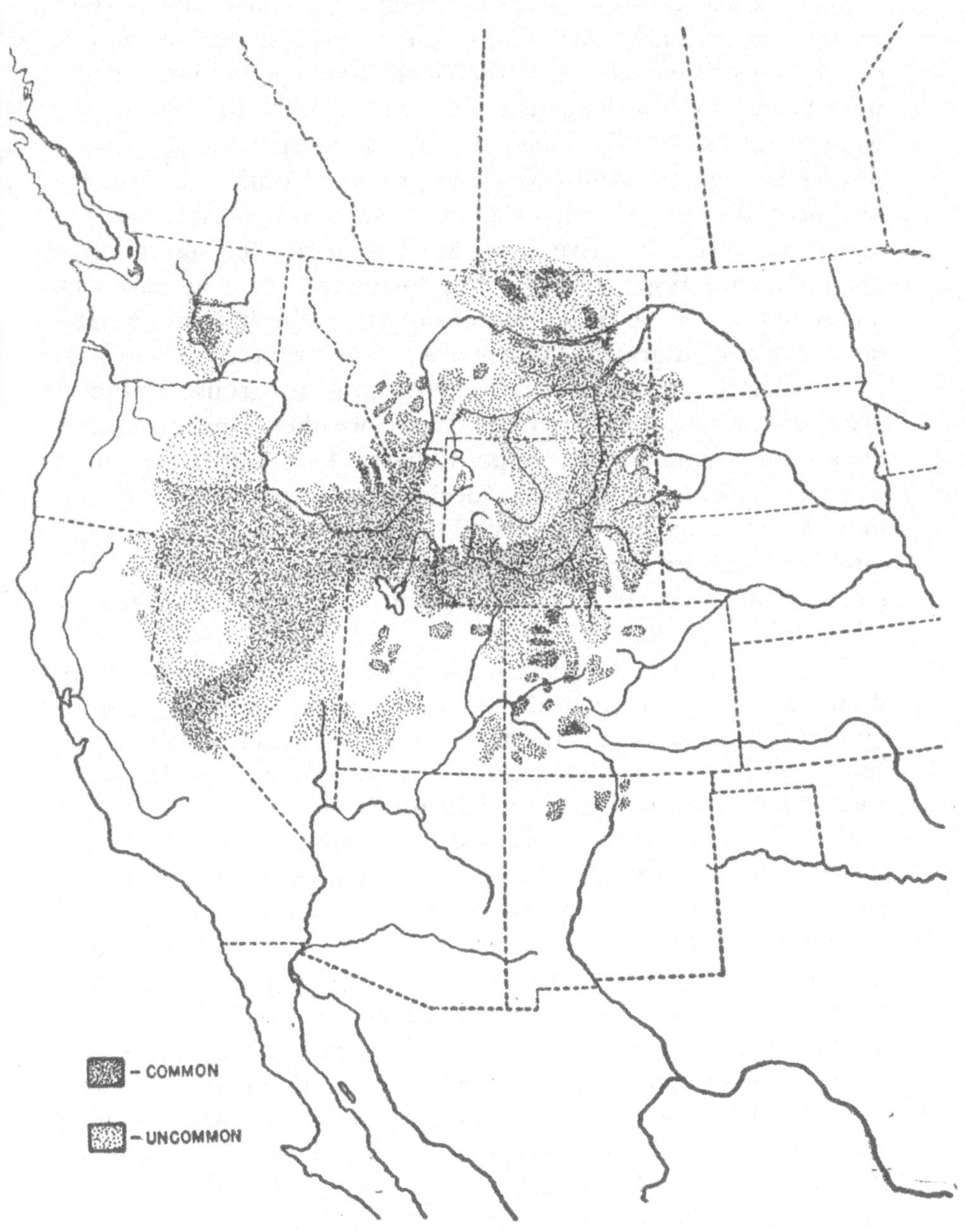

FIGURE 1.   PRESENT RANGE OF THE SAGE GROUSE

Most of the information concerning sage grouse populations during the period of decline and the causes responsible for this decline have been obtained from incidental observations of naturalists and wildlife biologists working on other projects in the western states. At the time of decline no specific population studies of the sage grouse were in progress. With a knowledge of recent land-use patterns and life-history relationships of the sage grouse, we can arrive at certain logical conclusions concerning the factors responsible for the downward trend in numbers. There appears to be little question that the elimination of habitat, uncontrolled grazing, and poorly devised protective measures were primarily responsible for the decline in grouse numbers prior to the mid-thirties. Unfavorable weather, predation, and disease were undoubtedly responsible for local fluctuations and may have operated as indirect decimating agents following periods of cyclic high populations. There is some evidence to indicate that sage grouse populations may have fluctuated in the same rhythmic patterns of scarcity and abundance exhibited by other grouse species.

Reports received in 1949 from western game departments have shown a wide variance in the actual years at which sage grouse numbers began their downward trend. This inconsistency in decline may be accounted for by the varying dates of reclamation and settlement of wilderness sagebrush areas.

New Mexico, at one time, had fairly large numbers of sage grouse in the counties of Taos, Rio Arribal, Sandoval, and San Juan, but by 1905 grouse in New Mexico were exterminated and then placed on the protected list! The species has apparently been exterminated in British Columbia, although the date has not been established. Washington experienced a steady decline beginning in the latter part of the nineteenth century, continuing until a low point was reached about 1935. Idaho reported steadily decreasing numbers of sage grouse until 1947. Nevada has noted several downward trends, the latest of which began in 1942 and ceased in 1945. According to studies in Oregon reported by Batterson and Morse (1948), no accurate measurements of population density were made prior to 1941. However, general observations and reports of Oregon residents indicate there has been a decline since 1919. Population studies made between 1941 and 1946 have recorded a decrease of 77 per cent on Baker County, Oregon study areas. However, the population curve in Oregon started upward in 1947, and continued to rise during 1948. Montana populations started to decline about 1943, and continued downward for a period of two years.

In order to trace the progress of the widespread upward trend reported in sage grouse populations throughout the West in 1948, the state and provincial game departments were again contacted in 1951. With the exception of Montana, all these wildlife agencies reported that sage grouse numbers continued their upward trend through 1950.

Bergeson (1951) reported the following observations for a small study area 40 miles southeast of Helena in Meagher County, Montana:

> On 32 miles of grouse transects established in 1942 and repeated in July, 1946, 1948, 1949, and 1950, it was found that the number of birds per mile in 1948 was 2.06, increased in 1949 to 2.94, and decreased in 1950 to 2.47. Average brood size decreased during these same three years respectively: 5.7; 4.7; and 3.7. On an established 20 mile road census route, the average number of birds per mile remained static at 6.0 birds in 1948 and 1949, but decreased in 1950 to 3.45 birds. The 1950 hatching and rearing season was, without question, quite adverse as reflected by the nearly equal old-young ratios obtained from our sharptail and mountain grouse species checking stations.

From this data and general information received from wardens, Bergeson believed that the trend since 1948 for the state as a whole had been generally downward.

There is some doubt that this data from Montana necessarily reflects a general downward trend in numbers since 1948, particularly since the number of birds observed per mile of transect increased from 2.06 in 1948 to 2.47 in 1950. In the absence of data on nest survivals and ratio of hens bringing forth young, early season brood counts would inadequately represent initial reproduction success. Breeding season populations furnish the most reliable indices to the previous year's productivity and to yearly population trends. It is probable that populations in Montana did not necessarily keep increasing during the 1948 to 1950 period. There is a likelihood, however, that the peak of abundance may have been reached between 1940-50, in which case minor fluctuations could be expected in localized areas.

The slight to marked increases in sage grouse populations reported in Washington, Oregon, California, Nevada, Utah, Idaho, New Mexico, North Dakota, South Dakota, Saskatchewan, and Wyoming have been further substantiated by the fact that open hunting seasons were held in seven of these areas in 1950 and 1951 (Table 2).

After a period of many years, Washington opened a restricted area to hunting in Kittitas, Grant, and Douglas counties in 1950 and 1951. Oregon established open seasons in several southeastern counties in 1949, 1950, and 1951. Reports from Oregon stated that none of these seasons reduced populations; also that birds were still substantially under-harvested. California, likewise, established open seasons on sage grouse in 1950 and 1951 in Mono County.

After a closure of six years, Nevada held open seasons on sage hens in 1949, 1950, and 1951, with a bag limit of five birds, and reported excellent hunter success. Utah conducted its first open seasons since 1932, in 1950 and 1951, in limited areas on a special permit system. Open seasons have been regularly held in Idaho since 1948 with excellent hunter success and high kills.

During the period of complete protection afforded sage grouse between 1937 and 1948, populations in Wyoming have recovered from a seriously low point. Over extensive areas in the state numbers have risen to a level permitting general open seasons.

In 1948, Wyoming permitted a ten-day controlled season on a special permit basis to alleviate damage to alfalfa crops in the Eden Valley district. There was no open season in Wyoming in 1949, although populations would have supported a general open season in large areas of the state. The first general open season since 1936 was conducted in four southwestern counties of Wyoming in 1950. A similar season was held in 1951.

In summarizing all of these seasons it may be stated that bag limits have varied from one bird per season (Washington) to five per day (Nevada), depending upon the extent of area opened to hunting, relative abundance of birds, and proximity of hunting area to population centers. Hunting seasons have opened as early as the first of August (Wyoming) and have been delayed as late as October 8 (Washington). Hunting seasons have covered most dates between these extremes in other states.

Montana has authorized no open season on sage grouse since 1943. Bergeson stated that the main reason for the extended closure in recent years has been the reluctance of sportsmen and the game department to open seasons in areas having small populations which they felt might be subjected to heavy hunting pressure by reason of their proximity to larger towns and cities. Furthermore, the sportsmen in the more remote areas have consistently refused to allow an open season unless all of the counties supporting sage grouse populations opened simultaneously. It was Bergeson's opinion that Montana could stand short open seasons in many areas of the state, *provided* that the season commenced later than the traditional late August or early September opening. He was certain that such early seasons do irreparable damage to sage grouse populations through exclusive hunting of young birds, or at least through inclusion of young birds only in the hunters' bag, with the attendant discarding of mature birds.

Historical Accounts of Sage Grouse Abundance

The known history of the sage grouse generally is dated from the accounts by Lewis and Clark of a "Cock of the Plains," which they found in the Rocky Mountains, about the headwaters of the Missouri, and afterward more abundantly on the plains of the Columbia. Coues (1874) related that the bird was very plentiful in colonial times and soon became well-known to nearly all of the explorers who had visited the regions it inhabited. Coues said,

> I have met them in the Laramie Plains, on the upper waters of the North Platte, on Sweetwater River, the headwaters of the Green River, on Lewis's Fork of the Columbia, and on the Wind River, but nowhere so numerous as on the latter stream and its tributaries, where scores would be often seen in a mile's ride. The mountain deserts constitute their home.

TABLE 2

A Summary of Sage Grouse Hunting Seasons in the West
from 1940 to 1951

State	Year	Date of Season	Area Opened to Hunting	Bag Limit	Average Bag	Approximate Kill
Calif.	1944	Sept. 9-10	Mono & Inyo co.	4	----	
	1950	Sept. 1-2	Mono County	2	1.37	3,500
	1951	Sept. 1-2	Mono County	2	1.62	3,000
Colo.	1945	Sept. (1)	Entire state	3	----	----
Idaho	1942	July 1, 5, 12	Lincoln County	2	----	----
	1943	Aug. 29, 31 and Sept. 2, 4, 5	Fifteen counties	3	----	----
	1948	Sept. 4-5	Eighteen counties	2	----	35,000
	1949	Sept. 17-18	Nineteen counties	2	----	40,000
	1950	Sept. 16-17	Twenty-four co.	2	1.09	30,000
	1951	Sept. 16	Twentyfive co.	2	1.04	17,500
Mont.	1940	Sept. 1-3	Forty counties	3	----	----
	1941	Aug. 17-19	Twenty counties	3	----	----
	1942	Aug. 23-25	Eighteen counties	3	----	----
	1943	Aug. 22-24	Twenty counties	3	----	----
Nev.	1940	Aug. 4-5	Northern 2/3 state	5	----	----
	1941	August	Northern 2/3 state	5	----	----
	1942	August	Northern 2/3 state	5	----	----
	1949	Aug. 21-22	Northern 2/3 state	5	4.0	20,000
	1950	Aug. 13	Northern 2/3 state	5	4.2	17,000
	1951	Aug. 19	Five counties	5	3.5	18,000
Ore.	1949	Oct. 1-5	Harney, Maleur co.	2*	----	500
	1950	Sept. 1-7	Deschutes, Harney, Maleur & Lake co.	3*	----	1,200
	1951	Aug. 25-Sept. 3	Similar to 1950	4	3.1	5,000
Utah	1950	Sept. 30-Oct. 1	350 permits on 6 units	2*	1.68	600
	1951	Sept. 15-16	1,120 permits on 14 units	4	2.93	2,500
Wash.	1950	Oct. 8-9	Kittitas County	1*	----	1,500
	1951	Oct. 7-8	Kittitas, Grant & Douglas counties	1*	----	1,200
Wyo.	1948	Aug. 1-10	1,000 permits in Eden Valley	6	4.55	2,500
	1950	Aug. 27-28 Sept. 2-3 & 9-10	Sublette, Uinta, Sweetwater, and Lincoln counties	3	2.7	22,000
	1951	Aug. 18-19	Similar to 1950	3	2.4	25,000

*Per season

Colonel John C. Fremont (1845) observed that the prairie hen (*Centrocercus urophasianus*) was still very abundant on the upper waters of the Green River in 1843. He further related that the *coq de prairie* was occasionally seen among the sage in the valley of the Grand River in Colorado.

Very few reliable estimates of the former abundance of sage grouse have been recorded. In the reports of old time residents, such expressions as "flocks that blackened the sky" are commonly used. A statement made by G. B. Grinnell (Bent, 1932) has been quoted to give an indication of the numbers of sage grouse which were encountered in western Wyoming:

> In western Wyoming the sage grouse packs in September and October. In October, 1886, when camped just below a high bluff on the border of Bates Hole, in Wyoming (south of Casper), I saw great numbers of these birds just after sunrise flying over my camp to the little spring which oozed out of the bluff 200 yards away. Looking up from the tent at the edge of the bluff above us, we could see projecting over it the heads of hundreds of birds, and as those standing there took flight, others stepped forward to occupy their places. The number of grouse which flew over the camp reminded me of the old time flights of passenger pigeons that I used to see when a boy. Before long the narrow valley where the water was, was a moving mass of gray. I have no means whatever of estimating the number of birds which I saw, but there must have been thousands of them.

After participating in shooting expeditions in Utah, Wyoming, and Montana, Huntington (1897) considered sage chickens to be most abundant in the vicinity of Fort Bridger, Wyoming, and southward to the Uinta mountains (Green River Basin).

Burnett (1905) listed the counties of Albany, Converse, Natrona, and Carbon as localities where sage grouse were most abundant in Wyoming. A single hunter has been known to kill a hundred birds a day without the aid of a dog.

According to a report by Visher (1913), sage grouse in South Dakota were restricted to Harding and Butte counties. He predicted that within a few years the birds would only occur in the state, as rare winter stragglers from Montana. He further implied

that homesteading was eliminating sage grouse habitat, and
was threatening to exterminate the birds.

Bailey (1925) related,

> Sage hens are still numerous in northwestern Colorado.
> They breed in all favorable places in Routt and Moffat coun-
> ties and then assemble in large flocks during the winter
> months. I have been told that their center of abundance dur-
> ing winter is in the general vicinity of Craig and Sunbeam,
> where they congregate on the great sage flats. During this time
> they are often entirely absent in places where they were com-
> mon in the summer.

He reported that sage hens were holding their own due to
closed seasons in Colorado and the inaccessibility of the regions
they inhabit. An open season was contemplated at that time
(1925); however, Bailey stated that he would be sorry to see this
last stronghold of the sage hen in Colorado invaded by hunters.

The only region in Canada where the sage hen existed in
1930 was in the valley of Frenchman's River in extreme southern
Saskatchewan (Brooks, 1930).

A Summary of Previous Studies on the Sage Grouse

Natural History Observations

Few American game birds have received less attention from
wildlife investigators, or have been the subject of more myth,
than the sage grouse. It has only been within comparatively recent years that preliminary life-history studies were undertaken,
and even these investigations were limited in scope and duration.
Sage chickens were abundant, widely distributed, and heavily
hunted throughout western United States and probably Canada;
it is amazing that their biological relationships were so little
understood by the masses of people, including some of the foremost naturalists of our time.

One of the earliest and most accurate accounts of the sage
grouse is presented by Coues (op. cit.) in which he gives a careful description of plumage, weights, measurements, food habits,
game and edible qualities, roosting, eggs, and some records of
former abundance.

Numerous references on brief life-history observations, geographical distribution, and occurrence of sage grouse have appeared from time to time in the various ornithological journals, local scientific publications, and natural history magazines. Many of these early accounts furnished valuable data on life-history stages and pointed out the conflict between sage grouse populations and advancing civilization. Other observers spread fantastic tales regarding some life processes of the bird, many of which have persisted to this day.

Bond (1900) explained with great detail that the breast feathers of the male in the breeding plumage were worn down as a result of the bird sliding along the ground on its gular pouch and breast, and presented a beautiful line drawing to substantiate his theory. It was not until thirty years later that Allan Brooks (op. cit.) refuted Bond's theory by demonstrating that the stiff breast feathers of the strutting male in reality are specialized feathers and are not derived in the manner described so graphically by Bond.

It is still generally believed that the sage grouse does not possess a gizzard. Coues (1874) quoted a statement by Ridgway,

> A peculiarity of this species, which I have not seen noticed, is that its stomach, instead of being hard and very muscular, as in other *Gallinacea*, is soft and membranous. This was first told to me by hunters in Nevada, and I afterwards satisfied myself of the truth of their statement that the sage hen "has no gizzard," by dissecting a sufficient number of individuals.

Horsfall (1932) reiterated this belief by stating that sage grouse have no muscular, grinding machinery known as a gizzard and that a daily requirement of the birds' diet is alkali mud or alkali water as an aid in passing the sage leaves which move unchanged through the digestive tract, except for loss of chlorophyll.

It is true that the stomach of the sage grouse is not as hard and muscular as in other gallinaceous birds. However, it is a strong-walled organ, muscular in structure, and by definition must be classified as a gizzard.

No less authorities on natural history than Grinnell, Bryant, and Storer (1918) stated that the rigors of winter cause a scarcity of food and by spring most sage grouse are poor in flesh as well as shabby in plumage. For three consecutive years, this worker has found sage grouse to be in their finest condition, at their greatest weight, and in their most brilliant plumage at the onset of the breeding season.

The first photographs ever obtained of strutting sage grouse have been credited to Horsfall (op. cit.) and were taken at Klamath Lake, Oregon. A field crew of the Wyoming sage grouse survey supposedly was the first to actually observe the mating performance of sage grouse. Simon (1940) also witnessed these mating acts in Wyoming and presented the first detailed accounts of the strutting and mating performance. Prior to 1940 the unique, polygamous behavior of sage grouse on their breeding grounds had been little understood, or even suspected. There are many people, even today, who believe that the cocks deposit the sperm on the strutting grounds and the hens pick up the sperm. Few individuals, indeed, have witnessed the mating act of sage grouse. Many others believe that the bird does not mate. It was with a knowledge of all of these circumstances that Scott (1942) investigated the entire mating cycle of the sage grouse, with particular emphasis on the behavior of both sexes and the complicated patterns of social dominance exhibited by the cocks.

Recent Sage Grouse Studies in the Western States

Wyoming—The first comprehensive study of the sage grouse was made by Girard (1937) in Sublette County, Wyoming, where intensive field observations were carried on during the summer of 1934. The purpose of this study was to investigate the life history, habits, and food of the sage grouse, and to secure information that could be used by western game departments in formulating measures designed to increase or maintain the species in its present habitat.

Numerous life-history observations were made by Girard. Studies of thirty-three stomachs and crops collected during the summer months were used to represent the summer food habits of sage grouse in a single locality. The percentages of vegetable and animal foods were 88.5 and 11.5 per cent, respectively. The

Compositae family furnished 73 per cent of the total food of sage grouse for the period studied. Ants made up almost 10 per cent of the total food contents.

Girard presented a plan for management which involved setting aside key refuge areas throughout the state, and recommended that the hunting season should be closed for at least three years. The hunting season, when next opened, should be delayed until September in contrast to the early August seasons which had prevailed. He was of the opinion that early season hunting was one of the most important factors in the reduction of the sage grouse.

It is of interest to note that Girard's studies were instrumental in effecting a closed season on sage grouse in Wyoming and other western states for an extended period of years, and therefore materially assisted game departments in applying more stringent protective measures in an attempt to restore sage grouse populations in the West.

Utah—The first studies on the sage grouse in Utah were made by Griner (1939) over a two-year period from May, 1936 to May, 1938. The work was primarily centered in the Strawberry Lake area of Utah, although range investigations were extended to cover other areas of Utah and Idaho. The studies of Griner and Girard were independently pursued, and the results of their field investigations were submitted as requirements for advanced degrees at The Utah State Agricultural College and The University of Wyoming, respectively.

The purpose of Griner's investigations on the sage grouse was (1) to obtain life-history data on the species; (2) to study environmental factors influencing the distribution and numbers of the sage grouse; and (3) to attempt to outline a plan of management which might stabilize grouse numbers in areas where still present.

Extensive food habits studies by Griner based upon stomachs collected from May to October, inclusive, confirmed earlier reports that the sage grouse is primarily a vegetarian. Plant material made up almost 98 per cent of the adult diet, and of this amount, the sagebrush (*Artemisia tridentata*) group made up

77.5 per cent. By August, the immature birds were essentially on a diet of plant material, consisting largely of sagebrush. Griner emphasized that during the winter months the different species of *Artemisia* provided practically the entire diet.

In discussing range and habitat factors, Griner believed that the marked decrease in sage grouse abundance could be attributed to the following factors:

1. Destruction of the forage and depletion of available water by overgrazing and burning practices.
2. Settlement and cultivation of original sage grouse range.

He traced the chronology of the settlement of land west of the Rockies and the resultant development of the livestock industry. The boom years of the 1880's resulted in the first real depletion of the range forage. Griner pointed out that the conflict between woolgrowers and cattlemen further intensified the competition for use of the range, resulting in additional depletion of forage resources. The war years caused a marked increase in range use with the result that the stockmen, as well as the range, faced ultimate ruin. He noted that the vast changes produced in the native vegetation due to overgrazing, burning, and settlement have seriously affected the welfare of many wildlife species, especially the sage grouse. He concluded that the Taylor Grazing Act terminated the era of free use and destruction of the range and predicted that, through education and new land-use programs of the Forest Service, the Soil Conservation Service, and the Grazing Service, it would again be possible to restore western ranges to a condition of productiveness for both livestock and wildlife.

The material in Griner's thesis was summarized and presented at the Third North American Wildlife Conference and subsequently appeared in the transactions of that meeting (Rasmussen and Griner, op. cit.).

Colorado—Sage grouse investigations in Colorado were launched in 1939 by the Game and Fish Commission as a federal aid project, and continued through 1941. The field work was confined to the North Park and Craig areas and involved primarily strutting ground, nesting, and brood studies. The relationship of environmental factors such as food, cover, and water to sage grouse utilization and nesting was studied during the spring and summer periods.

Post-mortem examinations were conducted on a large number of sage grouse, either found dead or collected for scientific purposes. Considerable data on the occurrence of disease-producing organisms and their pathogenicity were presented. In order to determine the causes of sage grouse nest destruction in Colorado, traps were erected around dummy nests, poisoned eggs were placed in artificial nests, and hair samples were collected from bushes in the vicinity of destroyed nests. In this manner, badgers were determined to be the key predator on sage grouse nests.

The survey disclosed that the nesting density in North Park was approximately thirty-one nests per square mile on a 1,280-acre tract. Average densities of thirty-seven and forty-five birds per square mile were found on two census areas in North Park.

No evidence of widespread disease epidemics was found among sage grouse in Colorado.

The Colorado investigations resulted in a report, *Sage Grouse Survey in Colorado,* published by the Game and Fish Commission.

Oregon—Research on sage grouse in Oregon was likewise sponsored by the Game Commission (Batterson and Morse, op. cit.). The purpose of this study was to determine the factors limiting the distribution and abundance of birds in the state. Most of the field work was done in Baker County and covered the period from 1941 to 1946. A short technical bulletin, *Oregon Sage Grouse,* was prepared and published by the Oregon Game Commission.

Batterson and Morse have described the sage grouse and its habitat, and presented a summary of the bird's yearly activities. Counts of males on strutting grounds revealed a 77 per cent downward trend in numbers between 1941 and 1946. Populations have increased, however, since 1946.

Nesting studies carried on during several years of the investigation revealed a low nesting success which was interpreted by the authors to indicate that avian predation was the greatest limiting factor on sage grouse survival in Oregon. Intensive predator control experiments resulted in increased nesting success on the study plots.

Oregon investigated the possibilities of trapping and transplanting mature birds as a phase of sage grouse management. Several hundred birds were trapped and moved to new areas. It was their belief that these operations afforded a relatively cheap and simple method of restocking depleted sage grouse ranges.

One of the few successful attempts to rear sage grouse in captivity was accomplished in Oregon. A small artificial propagation experiment was conducted in 1942, and six sage grouse chicks were raised to the age of six weeks, then released. Artificial propagation was recommended as being neither economically feasible nor justified for sage grouse.

Predation and disease were ranked as factors causing the cyclic trends encountered in Oregon sage grouse populations. Recommended management practices included intensive predator control, continued protection, co-operation with land management agencies, and water development on antelope ranges.

Current Investigations and Management Practices

Numerous studies have been made in the western states on the life history, food habits, ecology, behavior, and habitat requirements of the sage grouse. There have been few studies which have emphasized management of the species in relation to present and future land-utilization patterns. However, limited suggestions for management have been presented in studies undertaken privately, as well as those sponsored by the state game departments. Within the past fifteen years investigations dealing specifically with sage grouse have been inaugurated in several western states, notably Utah, Colorado, Oregon, and Wyoming.

Developmental projects which have been established by state and federal wildlife agencies for sage grouse restoration and management have included the Charles Sheldon Refuge in northern Nevada, the Hart Mountain Refuge in southern Oregon, and Strawberry Valley Refuge in Utah. The state of Montana has acquired several bird refuges totaling approximately 100,000 acres. These comprise in part some excellent sage grouse habitat. Since 1941, the Montana wildlife restoration division has taken over the development and maintenance of several dozen reservoirs and stock watering ponds. On these areas a portion of the shoreline has been fenced to regulate livestock use. Considerable planting of food and cover species has also been done. Some of these reservoirs have proved to be quite attractive to sage grouse. In view of the current mania for eradicating sagebrush, Montana game officials have contemplated acquiring a sizeable acreage of sage grouse habitat.

Washington and New Mexico have established small refuge areas to be used jointly by antelope and sage grouse. The latter state has successfully reintroduced sage grouse with stock obtained from Wyoming and South Dakota. Idaho, Oregon, and California have sponsored water development projects of admittedly questionable value due to the lack of completion and supervision during the war years.

Live-trapping and transplanting programs have been major activities of sage grouse projects in Wyoming, Oregon, and Montana during the past ten years. South Dakota has also trapped small numbers of sage grouse. The trapping and transplanting program has been considered in Utah and is now in the active planning stage. Live-trapping and restoration programs in most states have been an outgrowth of the recent increase in sage grouse populations, and more specifically, populations which were a source of actual damage to agricultural crops.

There has been considerable variation in the attitudes and policies of western game departments toward sage grouse. This may be attributed to the smaller numbers of birds encountered on the periphery of their range, and to a feeling on the part of many departments that research studies sponsored by other states would prove equally applicable to their own needs. Furthermore, there has been a widespread belief that sage grouse populations were slowly declining throughout their range, and in spite of man's efforts would be eliminated from the wildlife scene, or at least would cease to have any further importance as a game species.

Several states such as Colorado, California, Washington, Utah, and presumably Oregon have maintained a rather passive attitude toward sage grouse as a game bird, although they have harvested surpluses existing in localized areas. All western states have expressed a desire to preserve and, if possible, extend the range of this grouse. New Mexico considered the species as desirable game, and predicted the eventual re-establishment of the bird over its former range within the state by means of protection, range restoration, and water development.

Montana, Nevada, Idaho, and Wyoming have classified the sage grouse as an important game species and have had populations in recent years which have permitted either localized or general open seasons.

Analyses of reports from game departments of the various western states indicate that sage grouse populations have now increased to levels considerably higher than existed ten to fifteen years ago. This has been substantiated by investigations just completed in Wyoming. Within the past fifteen years, the sage grouse has responded to protection and improved management of range resources by both state and federal land management agencies, and once again has attained a position of prominence as a game bird in the West.

The Sage Grouse in Wyoming

Habitat and Distribution

Wyoming has within its borders dry, semiarid lands covered mainly with sagebrush and saltbush and more humid lands covered primarily with forests. The total land area of the state is slightly over 62 million acres. Excluding national parks and national forests, more than 23 million acres of publicly-owned land fall into the semiarid category and as such furnish potential habitat for sage grouse. In addition, several million acres of privately-owned range lands support vegetative types which are utilized both by livestock and sage grouse. The pattern of land ownership in Wyoming is listed in Table 3.

Sage grouse are found in all of Wyoming's twenty-three counties. With the exception of the extreme southeastern portion, they are generally distributed throughout the nonforested regions

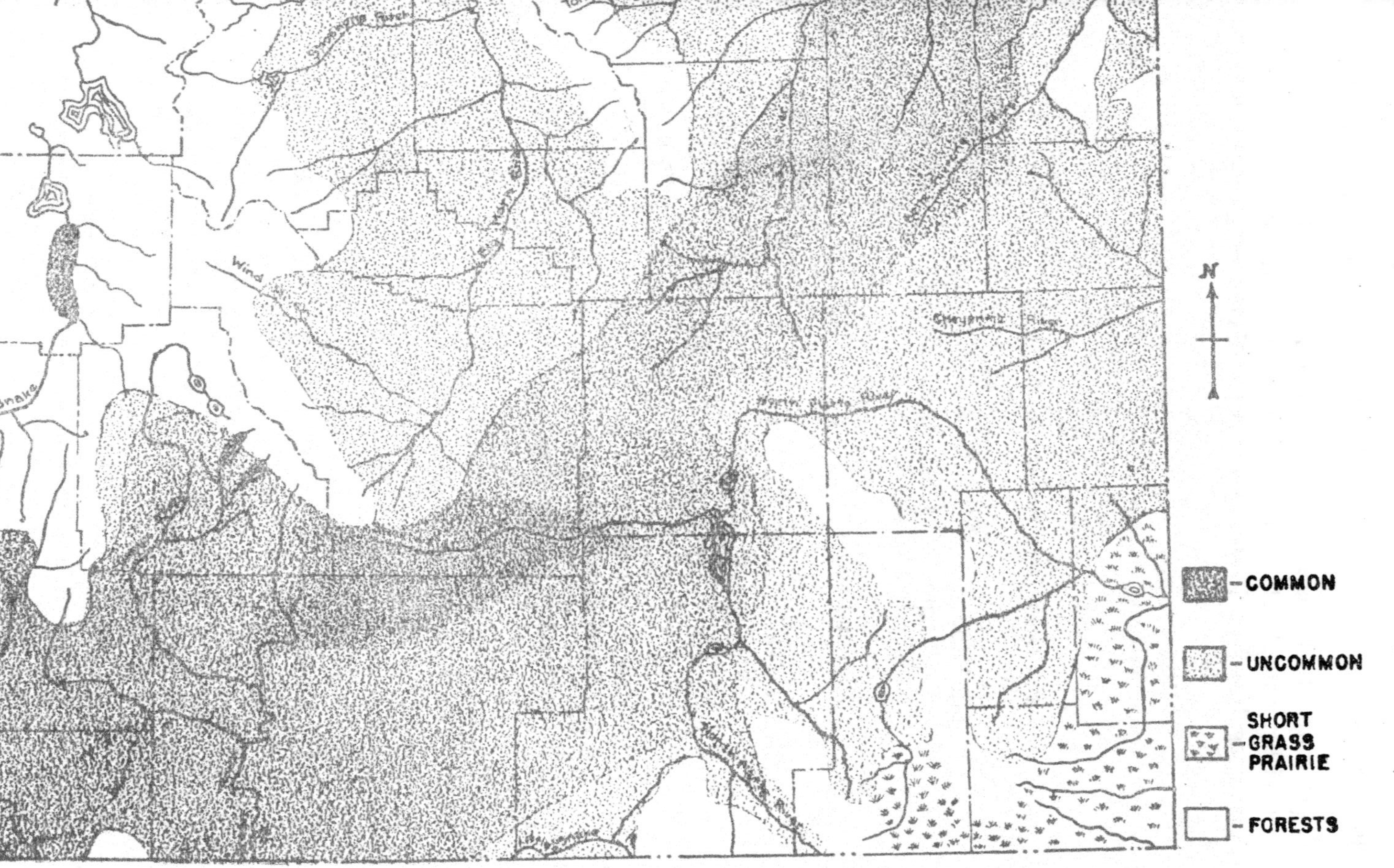

FIGURE 2. THE DISTRIBUTION OF SAGE GROUSE IN WYOMING

of the state. In areas heavily utilized for agriculture they have been either completely or partially displaced due to the elimination of sagebrush and associated plants. The heaviest concentrations of sage grouse in Wyoming are located on the upper drainages of the Green, North Platte, Sweetwater, Wind, and Powder river systems (Figure 2).

TABLE 3
*Land Ownership Pattern in Wyoming — 1950**

Type of Ownership	*Acreage*
Public Lands	
Taylor grazing lands	16,647,000
National forests	8,565,000
Indian lands	2,081,000
National parks	2,231,600
	29,524,600
State & County Lands	
School lands	4,503,000
County lands	259,000
	4,762,000
Privately Owned Lands	28,119,400
Total Land in Wyoming	62,406,000

*Data from Bureau of Agricultural Economics

Recent Population Trends and Present Numbers

In 1901, the Wyoming legislature enacted a law which classified all species of grouse as game birds and provided the Game and Fish Commission with power to protect these birds through the regulation of open and closed seasons. The first legal protection afforded sage grouse in Wyoming occurred in 1902, when the period from July 15 to October 15 was declared to be open for sage grouse hunting, there being no bag limit, however. A revision in the game bird law in 1909 set the number of game birds to be killed or in possession at eighteen, of which not more than twelve could be sage hens.

The annual reports of the Wyoming Game and Fish Commission furnish quite accurate accounts of the trends exhibited by sage grouse populations in Wyoming since 1900. The following statements relating to sage grouse abundance have been taken from these annual reports and represent rather accurate summaries of the knowledge and ideas of the men engaged in protecting and administering the wildlife resources of Wyoming during the years specified. Many of these ideas have been found to be extraordinarily sound in the light of several decades of experience. On the other hand, the viewpoints favoring early seasons as expressed by some of these men have undoubtedly fostered the unsound theories now seriously threatening the proper management of sage grouse in Wyoming, as well as in certain other states.

1906 — Special Assistant John B. Duncan, Sheridan — "Sage hens are decreasing rapidly. I would recommend a closed season for these birds."

Special Warden James S. Simpson, Grovont—"The bird law is all right, except that the sage hen season opens too late. The young birds can't be distinguished from the old ones."

1908 — State Game Warden D. C. Knowlin—"A few localities report an increase of sage hens, but taking the state as a whole, these birds are certainly on the decrease."

1909 — Season closed on sage hens in Sheridan and Natrona counties from 1909-1915.

1910 — Special Assistant C. D. Blaine, Cody — "Sage hens doomed to extinction in vicinity of Cody, owing to rapid settlement."

1914 — State Game Warden W. H. Seebohm—"I recommend that the open season on sage hens in Albany, Carbon, Laramie, and Sweetwater counties be from July 1 to August 31, instead of from July 15 to August 31, as now provided. Also, that the open season on these birds in the other counties be from July 15 to August 31, instead of from August 1 to August 31. I make this suggestion for the reason that as the law now stands, these birds are too large when the season opens and it is hard to determine the old birds from the young birds."

1916 — W. H. Seebohm—"In my opinion it would not be a bad

idea to close the season on sage hens in a few counties, and cut the bag limit in others. We find the automobile to be the source of much hunting in districts which heretofore were scarcely visited by anyone and in a very short time this state of affairs will be disastrous to the sage hen." (Seebohm recommended a bag limit of four sage hens per day).

1922 — State Game and Fish Commissioner Bruce Knowlin—"I recommend that the season on sage chickens be closed for a period of two years."

1926—State Game and Fish Commissioner A. A. Sanders—"I contend that it is high time we are doing something to conserve the bird that offers recreational sport to thousands of citizens each year. I also believe that a five-day open season should be adopted."

1930 — Bruce Knowlin—"In 1929 the Game and Fish Commission allowed seven days hunting on sage chickens and this last year three days."

1932 — State Game and Fish Commissioner Robert A. Hocker—"We have complaints from ranchers for damages to crops due to sage grouse."

1937 — Sage grouse were given complete protection throughout the state for the first time in the history of the Game and Fish Commission. The season was closed and no legal hunting allowed in 1937. The season remained closed throughout the state until 1948.

1940 — State Game Warden Lester Bagley—"The sage grouse has been protected by closed seasons for four years and has increased until some areas are over populated."

1942 — Lester Bagley—"The Commission is gratified over reports that the numbers of sage grouse have been increasing steadily in most parts of the state and it is possible that one or more counties may be opened to a limited sage grouse hunt next year."

1944 — Lester Bagley—"The great native sage hen or sage grouse has reached a very low point in population during the last three to five years. In spite of a closed season during the past seven years, the population does not seem to increase in most parts of the state. Rather large populations of sage grouse are found in Sublette

> and Carbon counties, but in the remainder of the state populations remain at a very low point."

1950 — Wyoming's first extensive open season on sage grouse since 1936 was conducted this year in the four southwestern counties of the state.

Briefly, this has been the history of sage grouse populations in Wyoming during the first half of the twentieth century as recorded in the annals of the Wyoming Game and Fish Commission. The patterns of civilization and land use which accompanied the fluctuations in Wyoming sage grouse populations during this period will be traced in later chapters.

The reports of the Game and Fish Commission indicate that sage grouse numbers in Wyoming dropped steadily until the middle 1930's. Twenty years prior to 1937, Wyoming's game protectors and administrators issued a clear call for action on behalf of the species, recommending closed seasons, reduced bag limits, and refuge areas. There was no recognition, however, of the irreparable damage being done to sage grouse populations through the continued policy of hunting during the July and August rearing period. In several states this policy has continued to jeopardize good sage grouse management. Spring shooting long ago disappeared from the scheme of modern waterfowl management. Summer shooting of sage chickens offers equally poor chances for the sustained production of harvestable quantities of birds.

Historical accounts of sage grouse numbers in Wyoming have revealed definite patterns of fluctuation, irrespective of overall trends. Complaints of damage and excellent hunting conditions indicated generally higher sage grouse populations in the early 1930's (1931-33). The influences of heavy shooting, early hunting seasons, overgrazing, and drought apparently had so reduced their numbers that the season was closed throughout the state in 1937.

During the years from 1937 to 1940, sage hens showed a pronounced increase, and some areas in the state were considered to be overpopulated by 1940. Certain sections of the state reported concentrations which would have justified local open seasons.

Damage to crops was again prevalent in agricultural areas in Eden Valley and in Sublette County. Live-trapping programs were initiated in these areas in 1940 and were productive of results until interrupted in 1943.

The late war years interfered with sage grouse studies in Wyoming. There was little public demand for sage grouse seasons during the mid-1940's. Compared with the success experienced in live-trapping from 1940 to 1943 (2,293 birds), and from 1949 to 1950 (3,196 birds), relatively few grouse were trapped in agricultural areas during 1946-48 (609 birds). In 1949-50, damage claims again increased sharply in Eden Valley and in a few other agricultural areas of the state. Excessively high grouse populations led to intensive live-trapping operations in 1949 and 1950, and general open seasons were held in 1950 and 1951.

The majority of states within the range of sage grouse have reported similar population trends within the past ten years. Numbers declined from the level existing in the early 1940's and reached a low point between 1945 and 1947. Oregon, for example, observed a decrease of 77 per cent in numbers on one study tract during the period from 1941 to 1946. The trend was upward, however, in 1947, and has continued to rise through 1950. Idaho, Montana, and Nevada have indicated strikingly similar periods of scarcity and abundance. Washington and Wyoming populations reached low points in 1935-36, and again in the mid-40's.

Allowing for variations of one to three years in the occurrence of these periods of abundance and scarcity, the writer has recognized "well-defined lows" from 1935 to 1937, and again from 1945 to 1947. Populations recovered from both of these "lows" to produce "well-defined highs" about 1941 and 1950. As pointed out by Grange (1949), the recognition of the exact year of the cyclic high or low is one of the most difficult of field observations. He has established the occurrence of a cycle among four species of grouse in Wisconsin, and furthermore has defined the regularity of this cycle. For a period of ninety years and for ten consecutive lows, Grange presented evidence which indicated that this grouse cycle has occurred on a regular ten-year amplitude. The records have indicated that this cycle, described by Grange, has occurred once in each decade, with a low point of abundance occurring in that year of each decade which ends with the numeral *seven* (with but two exceptions of one year displacement). During the last four cycles, the grouse have been at abundance levels in years that end with the numeral *one*. In making use of this oversimplified formula for dating the cycle, Grange reminds us that the recent low actually extended from 1944 to 1948 or 1949, depending upon the criterion of scarcity used; similarly, the high of 1951 will extend approximately from 1949 to 1953, provided the cycle repeats on the same schedule in the forthcoming high.

Considering the amount of reliable information available on the scarcity and abundance of sage grouse populations, it would seem highly probable that this species of grouse may also be exhibiting the cyclic phenomena or oscillations so well associated with most other grouse and upland game birds. In the case of each recognized high and low, there has been a startling resemblance to the mechanics of the grouse cycle, both in periodicity and regularity of occurrence. In the light of evidence now at hand, it would appear that the amplitude of the sage grouse cycle has been increasing due to protection and to the effects of improved land-management policies on the vast public grazing domain (sagebrush-grass and salt-desert shrub range types).

Investigations and Surveys in Wyoming Prior to 1948

Investigations dealing specifically with sage grouse were first sponsored by the Wyoming Game and Fish Commission in December, 1939. At this time a sage grouse survey was activated as a field project under the provisions of the newly enacted Federal Aid in Wildlife Restoration Act. The objectives of this survey as listed in the original project outline were as follows:

1. To map present distribution and abundance of sage grouse; to map former distribution and abundance; to map potential habitat; and to determine the extent of coccidiosis in sage grouse in Wyoming in recent years.

2. To study depredations on crops by sage grouse.

   3. To gather information on other game species during the
      course of the survey.
   4. To accomplish the redistribution of the species.

Work progressed on all phases of this project during the period 1940-43. The survey was interrupted in January, 1944, due
to personnel problems associated with the war years.

Information obtained from the report submitted by the project leader at the termination of the preliminary project has
indicated that the survey contributed results of questionable
value and application.

The investigations relating to coccidiosis were primarily concerned with the collection of samples of sage grouse droppings
on a statewide basis. Analyses of these droppings for the presence of coccidial oocysts revealed that carriers of coccidia were
present in almost every area where sage grouse occurred. However, only a few areas were known in which sage grouse suffered
heavy losses from the disease. It was during this period that
Ralph Honess, head of the Parasitology Department, University
of Wyoming, assumed charge of the investigations relating to
coccidiosis. These studies have continued in recent years as a
phase of the work of the Disease Research Laboratory under the
direction of Honess and sponsored jointly by the Wyoming
Game and Fish Commission and the University of Wyoming.

Considerable field work was done on the sage grouse breeding
cycle. Yearly nesting studies conducted at Pacific Springs, Eden
Valley, and at Big Piney (all in the Green River Basin) revealed
an excessively high percentage of nest destruction by predatory
birds. Survey personnel assumed that these high annual nesting
losses from predators were responsible for the gradual decline in
sage grouse numbers during the past half century. The Commission thereupon proposed that sportsmen's groups inaugurate extensive programs of predatory bird control. A predatory bird
control program consequently was launched in 1942. It was
stimulated by the publicity afforded the destruction of sage
grouse nests by predators and the inability of field personnel to
properly interpret predation and its influence upon wildlife
populations. Personnel on the sage grouse survey provided assist-

ance to sportsmen's clubs, ranchers, and other agencies in organizing programs for magpie and crow control. The current predatory bird control program, which has been operating in cooperation with the various counties on a 50-50 cost-sharing basis, was an outgrowth of the early sage grouse nesting studies. The Game and Fish Commission matches dollar for dollar the funds expended by private agencies for predatory bird and skunk control.

Nesting studies conducted from 1948 to 1950 have continued to reveal high losses due to nest destruction. Throughout the last twenty years sage grouse populations have periodically declined, only to increase in some areas of the West to levels almost precluding an adequate harvest of surplus numbers. During these periods of scarcity and abundance, game and fish funds have been diverted to unwise and uneconomical programs of predatory bird control.

As noted previously, a sage grouse live-trapping and transplanting program was inaugurated in Wyoming in 1940 for the purpose of redistributing concentrations of birds. Field crews encountered excellent trapping success considering that the project was in the experimental stages. During the four years from 1940 to 1943, nearly 2,300 birds were trapped in Eden Valley and adjacent areas, and transplanted into other regions of the state. According to reports issued in 1943, preliminary surveys of the transplant areas indicated that the liberated birds had remained in the vicinity of release, although no definite information was obtained on their breeding success. Trapping operations produced high catches of grouse, but the success of the transplanting program must be re-evaluated in the light of information gathered from the movements of banded birds released during similar operations conducted from 1948 to 1950.

Supplemental studies included in the initial project resulted in three small publications relating to sage grouse. Ward et al. (1942) presented the results of tests on the susceptibility and effects of strychnine poisoning on sage grouse. The structure and function of the neck muscles of sage grouse as related to breeding condition of the bird have been summarized by Honess and

Allred (1942). Appearing with the latter paper, in a joint bulletin, is the work of Clarke and associates (1942) on the variations by season and sex in the "air sacs" region of the sage grouse.

The sage grouse survey was reactivated in July, 1946, under different leadership, but continued to function in accordance with the scheme of organization and planning associated with the earlier project. Additional studies were made on strutting, mating, and nesting activities. These projects were not performed on a specific land area basis; on the contrary, the various phases were done at scattered, nonrelated localities: the strutting observations in Lincoln County (Kemmerer), the nesting studies in Fremont County (Pacific Springs), and the trapping operations in Sweetwater County (Eden Valley).

The trapping and transplanting program was renewed during the period 1946-48, but operated on a somewhat reduced scale due to the smaller number of sage grouse frequenting alfalfa fields.

The biennial aerial census of antelope during the winter of 1947-48 afforded excellent opportunities for survey crews to observe winter concentrations of sage grouse throughout the state. Large concentrations of flocks containing from several hundred to several thousand birds were noted in Johnson, Natrona, Sweetwater, Carbon, and Fremont counties.

An attempt has been made in the preceding discussion to summarize the scope of activities and accomplishments of the sage grouse survey in Wyoming during the years from 1939 to 1943 and from 1946 to 1948. Occasional references will be made to specific phases of that work as it may supplement or contribute new information to the investigations conducted during the final years of the project.

Nature and Scope of the Present Study

The upper Green River Basin of western Wyoming has long been famous as habitat for sage grouse. The Green River forms the main upper channel of the Colorado and was named by the Spaniards about 1817 (Fremont, op. cit.). In the narratives of his exploring expedition to the Rocky Mountains in the year 1842, and to Oregon and North California in the years 1843-44, Fremont recorded that it (Green River) was then familiarly known as the Seeds-ke-dee-agie, or Prairie Hen (*Tetrao urophasianus*) River—a name which it received from the Crow Indians, to whom its upper waters belonged and on which the sage grouse was still very abundant.

This valley which supported sage grouse populations so impressive to Indian tribes and white men has been little changed by civilization. It has survived today as one of the largest primitive sagebrush areas on the continent. This Basin was selected as the setting for the intensive studies on sage grouse which were begun in June, 1948 and concluded in 1951. For the purpose of the study, two adjacent areas were selected in northern Sweetwater, southeastern Sublette, and southwestern Fremont counties. The one tract (Figure 3) embraced 205 square miles of land within the Eden Valley irrigation district and was classified as agricultural land in the sense that it was either being farmed, or else was capable of being placed under irrigation at some future date. Actually only 9,000 acres were under cultivation during the period of study. The other tract (Figure 4) contained roughly 83 square miles, and included the territory lying between Dry Sandy Creek and Pacific Creek from their sources in the foothill region of the Wind River mountain range to their respective outlets into Eden Valley. With the exception of a state highway and an oil drilling location, the latter area was considered to be primitive semidesert sagebrush country.

The present study has been concerned primarily with yearlong investigations focused on an intimately known area of land. Emphasis has been placed upon land-use activities as related to the abundance, distribution, productivity, and management of sage grouse populations. The trapping and transplanting program was continued intensively in an attempt to obtain information on the seasonal movements of sage grouse, to evaluate the success of transplanting programs, and to reduce populations which were damaging alfalfa crops in Eden Valley.

The ecological and natural history aspects of sage grouse populations have been considered at length, but neither subject was developed as a complete monograph. Careful reviews have been made of the literature and previous investigations on sage grouse throughout its range. The 1950 and 1951 sage grouse hunting seasons in Wyoming were analyzed since they were localized in the Green River Basin, and the heaviest hunting was done in Eden Valley. Intensive strutting ground censuses, nesting studies, brood observations, and photographic work were conducted on the study tracts. Most of the field work has been confined to the study tracts, although observations on the seasonal movements of grouse were extended to include considerably larger areas in the Basin.

The primitive Jackson Hole district was utilized for experimental studies on dispersal and survival of sage grouse following their removal to restoration areas, and also as a control area for determining population trends.

The entire study has been supplemented and richly enhanced by accounts from other western game departments on the history and present status of sage grouse populations in their states.

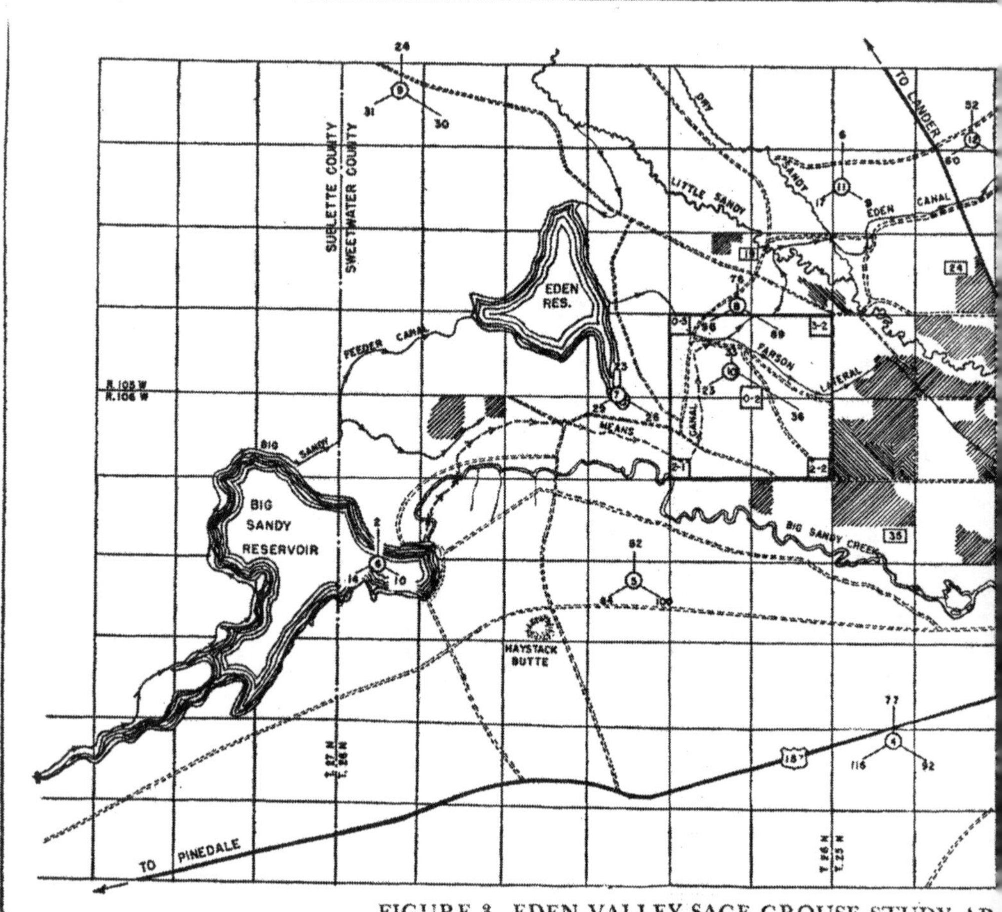

FIGURE 3. EDEN VALLEY SAGE GROUSE STUDY AR

(Adapted from Bureau of Reclamation and Soil Conservation Service Maps of Eden Vall

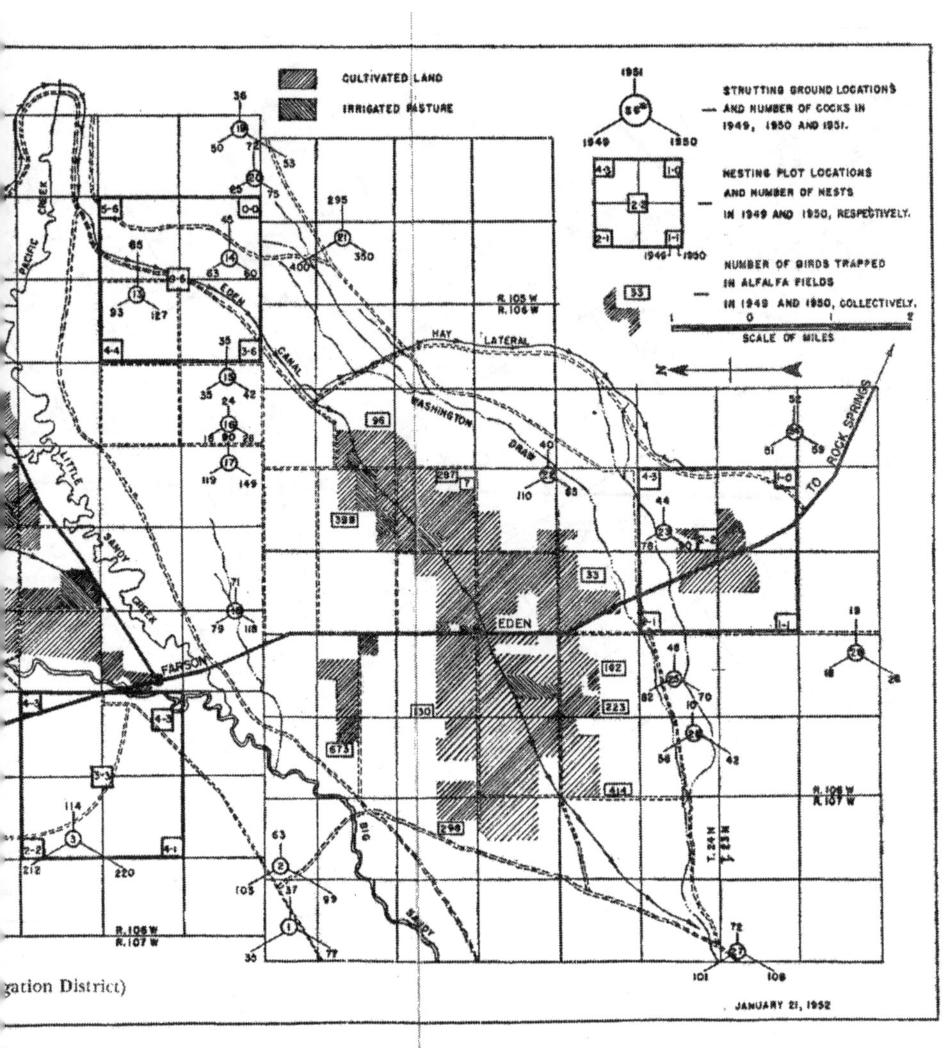

gation District)

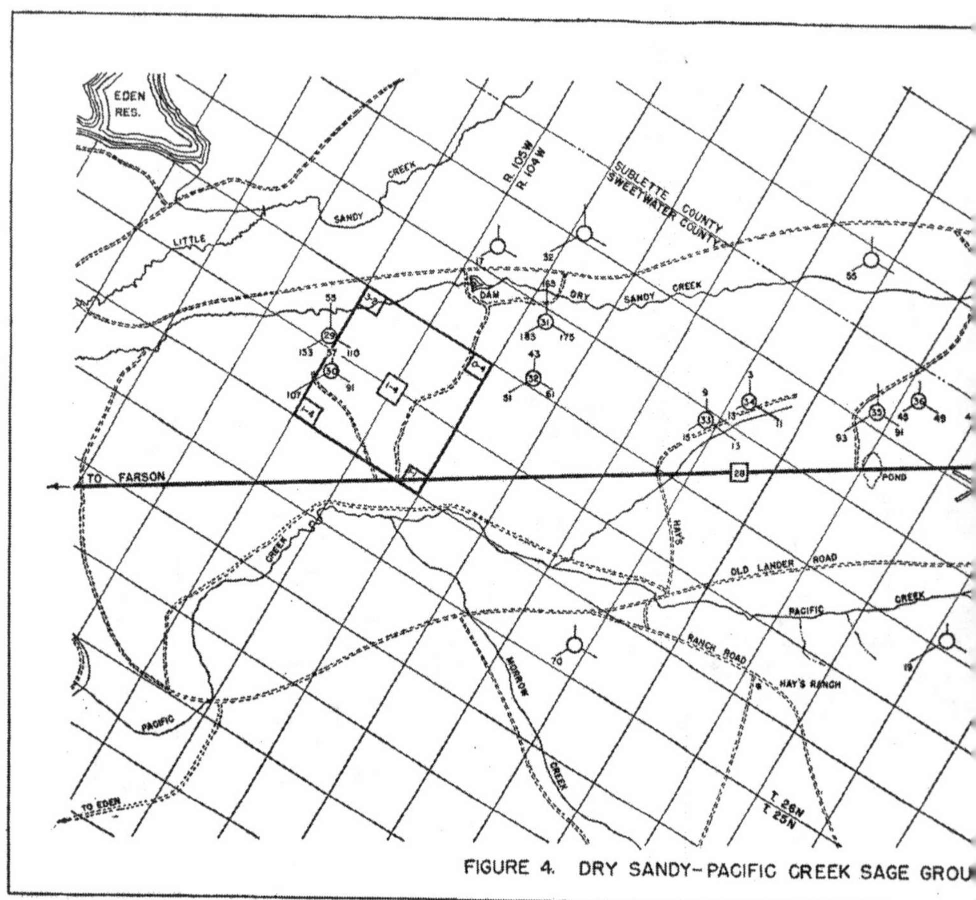

FIGURE 4. DRY SANDY-PACIFIC CREEK SAGE GROU

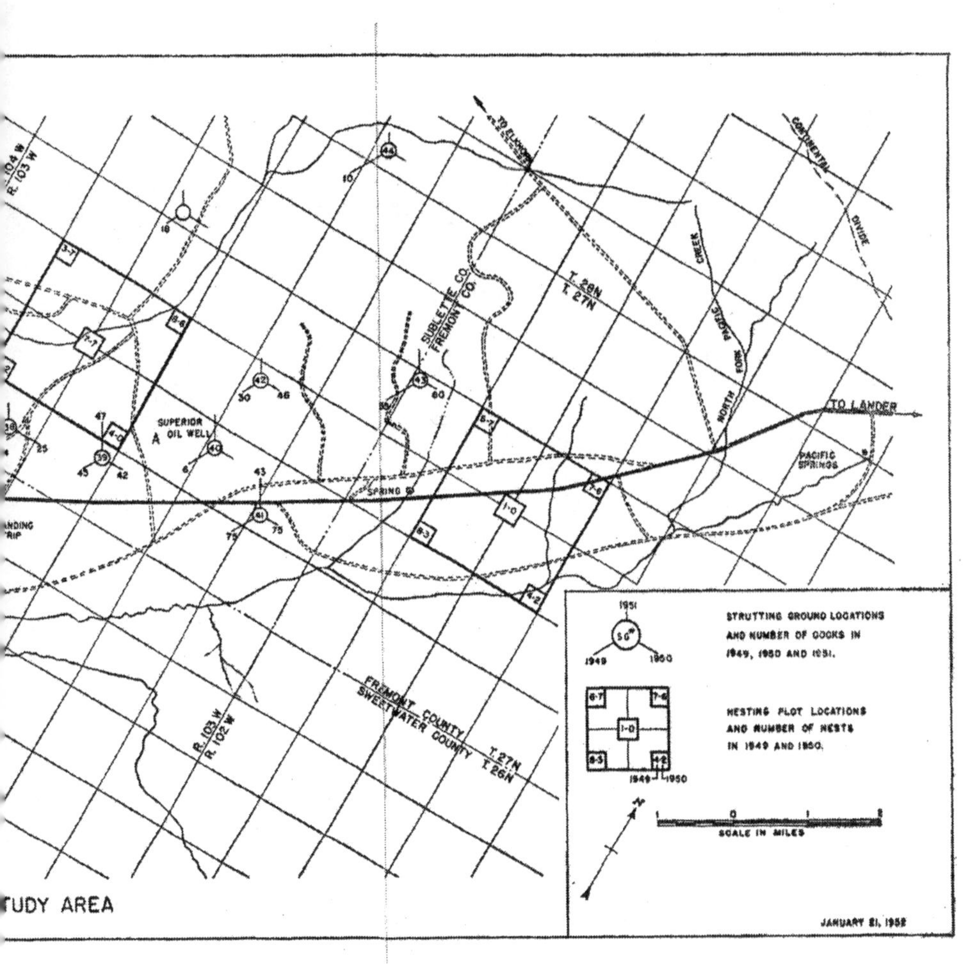

TO LANDER
TO LANDER
TO ELKHORN
CONTINENTAL DIVIDE
NORTH FORK
PACIFIC CREEK
PACIFIC SPRINGS
SUBLETTE CO.
FREMONT CO.
T. 28N
T. 27N
SUPERIOR A OIL WELL
SPRING
FREMONT COUNTY
SWEETWATER COUNTY
T. 27N
T. 26N
R. 104 W
R. 103 W
R. 103 W
R. 102 W
LANDING STRIP
STUDY AREA
1951
SG
1949
1950
STRUTTING GROUND LOCATIONS
AND NUMBER OF COCKS IN
1949, 1950 AND 1951.
6-7
7-6
1-0
8-3
4-2
1949
1950
NESTING PLOT LOCATIONS
AND NUMBER OF NESTS
IN 1949 AND 1950.
N
0   1   2
SCALE IN MILES
JANUARY 21, 1952

THE SAGE GROUSE AND ITS ENVIRONMENT
IN THE UPPER GREEN RIVER BASIN

Physical Environment

Geological Features of the Landscape

The geology of the Green River Basin has been influenced by the Uinta, Wind River, and Wyoming mountain ranges which surround the Basin. The Uintas form the southern boundary of the Basin and extend from the Wasatch Range eastward across the southern part of Wyoming into northwestern Colorado. The Wyoming Range flanks the Basin on the western side, and the Wind River Range bounds the tract to the north and east. This river basin is filled with gently warped Tertiary sediments, mostly of Eocene age. These sediments lap upon the flanks of the surrounding mountains from which they were chiefly derived. The deposits are known as the Wasatch, Green River, and Bridger formations.

The Eden Valley irrigation district is underlain by the Bridger and Green River geological formations. The Bridger underlies the northern or Farson segment of the district, and the Green

River, the older of the two, underlies the southern or Eden segment of the district. In general, the northeast-southwest contact of these two formations may be located in the vicinity of the Little Sandy and Pacific creeks which serve as a boundary between the Farson and Eden segments of the district.

Topography of the Basin

The Green River originates in Wyoming at the northern extremity of the Wind River Range at an elevation of nearly 11,000 feet, from whence it flows in a southerly direction approximately 480 miles to empty into the Colorado River in Utah. The elevation at the mouth of the Green River is about 4,000 feet. The headwater region is characterized by high mountains. The Basin itself is distinguished by flat to rolling sagebrush hills dominated by generally rugged buttes and mesas. Deep draws and arroyos feature the lower drainages of streams tributary to the Green River. Glacial outwash material is a prominent characteristic of bottom lands adjacent to the larger streams. One of the largest sand dune areas on the continent is located in the Basin, southeast of Eden Valley. The central portion of the Basin is known as the Little Colorado Desert, but topographically it differs very little from the rest of the area.

The Eden Valley irrigation district, at an elevation of about 6,600 feet, occupies a portion of the broad, flat valleys of Big Sandy and Little Sandy creeks, tributaries of the Green.

Soil Conditions

The soils in the upper Green River Valley are derived chiefly from the Bridger and Green River formations. The Bridger is composed of dark brown sandstone and varicolored shales of marine origin. The Green River deposits consist of thin sandstone and limestone beds with thick strata of gray shales, and are highly fossiliferous, containing fresh water species of fish.

The green- and brown-sandy shales of the Bridger predominate as the underlying rock strata within the Farson segment. The gray, fissile shale of the upper part of the Green River formation outcrops persistently in the Eden segment.

In so far as origin is concerned, the major soil types in the

Eden Valley area fall into three distinct groups: residual soils, weathered from parent rock and in place; alluvial soils, deposited in the area by water or stream action; and eolian soils, deposited in the area by wind action.

Residual soils cover an extensive part of the western half of the Eden Valley area. These soils retain the inherent characteristics of the underlying rock. In the Farson segment, the soils that have weathered from shales are dark in color, gray or green, and contain varying amounts of "alkali" salts.

Seep and associated saline conditions are very evident wherever fairly thin irrigated soils overlie Bridger shales. The shallow outwash and residual soils, underlain by the Green River shales, are seldom wet or saline. This striking contrast in soil conditions may be attributed to the physical nature of the underlying rock. Bridger shales are dense, saline, and relatively impermeable, and produce more or less saline-alkali soils; whereas Green River shales are less dense, relatively permeable, less saline, and produce soils which contain little or no salts.

Water Supply

With the exception of the semiarid central portion, surface water supplies are fairly abundant during all seasons of the year and are generally distributed throughout the Basin. Due to the character of the upper watershed, numerous small drainages discharge from the mountainous regions into the Green River proper. Runoff from melting snows usually exceeds the early season requirements for irrigation, livestock, and wildlife; although later in the season the stream flow is considerably reduced. At this time of the year, many smaller streams dry up completely as they discharge onto the semiarid desert. The scattered buttes and mesas give rise to free-flowing springs from snow terraces on their leeward slopes. Diversion canal systems which irrigate hay and grain crops in the foothills region frequently divert the bulk of the water supply from smaller creeks, thereby creating dry stream beds on the lower drainages. As a result of decreased water supply in late summer, wildlife populations tend to congregate around desert waterholes, permanent springs, and larger

TABLE 4

Annual Flow of Big Sandy and Little Sandy Creeks
from 1931 to 1947 in Eden Valley (a)

Year	Big Sandy Creek	Little Sandy Creek
1931	35.2 (b)	19.6
1932	61.8	16.3
1933	49.6	12.5
1934	31.0	8.1
1935	49.7	13.3
1936	80.9	19.1
1937	59.4	14.3
1938	66.5	17.3
1939	58.2	16.2
1940	29.0	7.7
1941	60.8	13.1
1942	62.1	15.4
1943	77.1	21.6
1944	60.0	15.7
1945	60.7	14.8
1946	60.8	16.1
1947	79.6	20.0
Total	982.4	251.1
Mean	57.8	14.8

(a) Data from Geological Survey and Bureau of Reclamation
(b) Unit = 1,000 acre-feet

streams. The range development programs sponsored by the
Bureau of Land Management, the Department of Agriculture,
and the livestock industry, have provided important sources of
supplemental water supplies in semiarid regions for wildlife and
livestock. During the winter months, snow cover furnishes the
moisture requirements for livestock and wildlife, especially on
the arid central part of the Basin.

The Eden Valley irrigation district water supply is furnished
by discharges from Big Sandy and Little Sandy creeks. Runoff
from both creeks is diverted into an offstream reservoir from
which radiates an irrigation system for the Farson and Eden segments of the district. The average annual stream flow of Big

Sandy and Little Sandy creeks within the irrigation district has been calculated at 72,600 acre-feet for the period from 1931 to 1947, inclusive. The drainage area for these two watersheds is 94 square miles. The average annual runoff for each of these streams is summarized in Table 4. It will be noted that the figures for 1931, 1933, 1934, and 1935 indicate a subnormal flow indicative of low precipitation and drought conditions of the early 1930's.

Dry Sandy and Pacific creeks, tributaries of Little Sandy Creek, serve as boundaries of the Dry Sandy—Pacific Creek study tract and transport runoff water from this area. Following the spring thaw, the lower stretches of both creeks are classified as intermittent streams. A series of natural springs and seeps provide important sources of supplemental water on the upper drainages of Dry Sandy and Pacific creeks.

There is a striking difference in ground water behavior in the localities referred to as the Farson and Eden segments.

In the Farson segment, the saline soils are related to the Bridger shales. Ground water, the source of which is loss from the irrigation system, is held in the soils above the shales, especially in areas where the slopes are flat and lateral movement of ground water is extremely slow. Since the soils and Bridger shales are saline, a continually saturated soil provides a medium for salts to rise upward by pressure and capillarity. The salts concentrate on the surface simultaneously with the evaporation of the water carrier.

The Eden segment has a smaller acreage of land that is wet and alkaline. The geology and soil conditions indicate that adequate internal drainage exists. Summarization of the soils, geology, and ground water conditions in the Eden segment may be set forth as follows: (1) residual soils derived from Green River shales are relatively free from salinity since the shales originally were of the fresh-water type, having been deposited in fresh water lakes or inland seas under humid or semihumid climatic conditions; (2) under good subsurface drainage conditions, the small percentage of soluble salts contained in the residual soils is leached out rapidly by normal precipitation; (3) the underlying

Green River shale is thin-bedded, hard, and severely fractured, containing myriads of crevices or fissures which give this rock a high coefficient of permeability, very probably equal to coarse sand; and (4) this fissile shale member is of sufficient thickness to absorb all the water that percolates into it from irrigation systems and precipitation. The water table during the irrigation season, therefore, remains at some depth in the shale rather than in the upper soil horizon.

As will be demonstrated later in the land-use section of this report, the character of the two major geological formations in Eden Valley exerts a marked effect upon sage grouse populations in the Eden and Farson segments.

Climate

A semiarid or steppe climate is typical of the central portion of the upper Green River Basin. The annual precipitation varies from 5 inches on the desert to 30 inches in the foothills region. The area is protected to the north and northeast by the Wind River Range and the Continental Divide, and to the west and south by the Wyoming and Uinta ranges. These mountain barriers protect the region from a large portion of the cold air masses moving down from Canada and, also, from those moving in from the west. Because of this protection, storms generally are shorter and less severe than those occurring farther east on the Great Plains. Nevertheless, the winters are long and the summers short, due mainly to the high elevation. As a result of aridity and high elevation, the diurnal and seasonal variation in temperature is quite large, ranging on the average 35° F. daily and 90° F. annually. The prevailing wind is west and the average velocity high (12 m.p.h. at Rock Springs), with the maximum velocity occurring during the hours of greatest heating. Strong winds tend to be beneficial to livestock and wildlife, since a heavy snowfall without wind may shut off the food supply.

Meteorological summaries for the period from 1908 to 1950 were obtained from U. S. Weather Bureau station records at Farson. They are applicable throughout the Eden Valley study area and a portion of the Dry Sandy—Pacific Creek region.

The warmest month of the year is July with an average monthly temperature of 64° F. and a maximum temperature of 97° F., which occurred in July, 1931. The coldest month is January with an average monthly temperature of 9.3° F., and a record low of -48° F. in January, 1922. The average annual temperature is 37.7° F. The shortest growing seasons were recorded in 1931 and 1932 when freezing temperatures and killing frosts were recorded in every month during those two years. The longest growing season was 125 days in 1935. Average length of growing season was 87 days. Average annual precipitation was 7.24 inches with the greatest amount falling in June. Seasonal precipitation was highest during the spring. The heaviest monthly snowfall was 18.8 inches, recorded in February, 1909. The average annual snowfall was 27.3 inches. Snow cover generally remained on the ground during the winter months. The average number of clear days per year was 207; partly cloudy, 88; and cloudy, 70. Daily precipitation amounted to 0.01 inches or more on an average of 48 days during the year.

Records of relative humidity were obtained from the Weather Bureau station at Rock Springs airport, 40 miles south of Eden Valley, at an elevation of 6,781 feet. The average monthly relative humidity (11:30 a.m.) varied from 68 per cent in December to 28 per cent in July.

Physical Environment and Sage Grouse Activity

The region is considered to be well-suited to occupancy and use by sage grouse due to the nature of the physical environment.

The physiography of the Green River Basin has exerted a marked influence upon the development of daily and seasonal activity patterns of sage grouse. Generally speaking, sage chickens move into the foothills during the spring and summer and retire to the semiarid desert during the fall and winter. These seasonal movements have developed primarily as a result of the wide variations in the nature, amount, and distribution of water, and the availability of sagebrush for food and shelter requirements. These variations have been caused by seasonal differences in climate influenced by elevations and the proximity of some land areas to the mountains. The composition and drainage features of major soil types have had an additional influence upon sage grouse activity due to the nature of their vegetative types and water; e.g., sandy, well-drained soils supported sagebrush-grass types and fresh water springs, whereas poorly-drained shale soils featured the salt-desert shrubs, alkaline flats, and saline water.

The combined effect of these influences has produced a powerfully mobile bird, capable of extensive flight in the search for food, shelter, and water, yet well-adapted for survival during long and severe winters.

Plant Environment

Vegetative Types

Sagebrush (*Artemisia tridentata*) is the most common and widely distributed shrub in western North America. It is especially abundant on the arid plains of the Great Basin, where it is the dominant plant of the sagebrush climax, ranging to timber line in the mountains. Throughout other areas of its present range it has been described as an invader, during recent geologic time, of the climax grassland, desert scrub, and woodland formations. The most frequent associates of sagebrush are subdominant shrubs such as shadscale (*Atriplex confertifolia*), rabbitbrush (*Chrysothamnus* sp.), winterfat (*Eurotia lanata*), silver sagebrush (*A. cana*), and bud sagebrush (*A. spinescens*).

In their monograph on the North American species of *Artemisia, Chrysothamnus*, and *Atriplex*, Hall and Clements (1923) have included within the *A. tridentata* group a phylogenetic series of seven closely related subspecies, namely: *A. t. typica, A. t. nova, A. t. arbuscula, A. t. parishi, A. t. trifida, A. t. rothrocki*, and *A. t. bolanderi*. The sagebrush type proper is frequently associated with one or more of its variads, such as black sagebrush (*A. t. nova*).

Numerous workers in the fields of forestry, range and wildlife management, and ecology have been inclined to refer to the climax sagebrush type as merely *Artemisia tridentata*, when actually their broad grouping included in most cases several of the subspecies of *A. tridentata*, as presented by Hall and Clements.

Herbarium specimens of the subspecies *typica, nova, arbuscula,* and possibly others, have recently been assigned to full specific rank by George H. Ward, Stanford University botanist, who is revising the *A. tridentata* complex. We have, then, two closely allied classification concepts regarding the *A. tridentata* group, depending upon the botanical viewpoint. Regardless of the interpretation, it is well to remember the close relationship exhibited by plants within this group.

Coulter and Nelson (1909) referred to *A. tridentata* as the black sagebrush, although in recent years most workers have commonly called this species big sagebrush and assigned the name black sagebrush to *Artemisia nova,* or variety *nova* as the case may be. Big sagebrush and black sagebrush occur commonly and form pure types throughout large areas of the intermountain and Great Basin regions. The former plant is an indicator of deep, well-drained soils of high water content, whereas the black sagebrush is more commonly associated with shallow soils and less productive sites. The individual growth form of each species varies, depending upon site factors and grazing conditions.

The sagebrush and grass climaxes evolved as a result of the uplift of the Rocky Mountain system some 20 or 30 million years ago. Although the annual precipitation over sagebrush areas may range from 5 inches to 30 inches, the average annual precipitation within the climax is less than 10 inches. This is not sufficient for a good growth of the sod-forming grasses but does favor the growth of palatable perennial bunch grasses such as western wheatgrass (*Agropyron smithii* Rydb.), needle and thread grass (*Stipa comata* Trin. and Rupr.), Indian ricegrass (*Oryzopsis hymenoides* Ricker), bluebunch wheatgrass (*A. spicatum* Scribn. and Smith), Sandberg bluegrass (*Poa secunda* Presl.), sand dropseed (*Sporobolus cryptandrus* A. Gray), and giant wild rye (*Elymus cinereus* Scribn. and Merr.).

The Red Desert lies in southwestern Wyoming and has been defined as that area situated between the North Platte and Green rivers, and south of the Sweetwater divide. It embraces almost seven million acres of public range lands. The Eden Valley and Dry Sandy—Pacific Creek study plots are located within this region. The principal plants of the Red Desert are the true sages (*Artemisia*), salt sages (*Atriplex*), white sage or winterfat (*Eurotia*), greasewood (*Sarcobatus*), rabbitbrush (*Chrysothamnus*), and the grasses previously listed. The most abundant plant in the region is *A. tridentata*.

Vass and Lang (1938) have conducted detailed studies of the vegetative composition of the Red Desert area. As a result of this work, they prepared a chart showing the proportion of Sweetwater County which is occupied by the various vegetative types (Figure 5). The boundaries of Sweetwater County (the county contains 6,715,000 acres) are roughly the boundaries of the Red Desert. The Eden Valley area lies in the northcentral portion of the county.

The vegetative chart indicates that large areas in the county supported dominant stands of the *A. tridentata* group. However, it usually occurred in combination with other subdominant plants. The sagebrush-wheatgrass type was the most common type found in Sweetwater County, covering almost one-third of its area.

Salt sage (*Atriplex nuttallii*) was commonly found in nearly pure stands and occurred as the dominant plant in eight major types. The salt sage type was usually confined to the long, narrow valleys and low, alkali flats. Greasewood types were generally encountered on flats and gulches heavily impregnated with alkali.

Rabbitbrush was uncommonly found as the dominant plant, except in the *A. tridentata* type. Numerous species of rabbitbrush were found in nearly all of the vegetative types of the county. They thrived on a variety of land forms and were apparently unaffected by varying conditions of moisture and salinity.

Thirty-eight plant families were encountered quite commonly by Vass and Lang in Sweetwater County. The composite (*Compositae*) family was represented by the largest number of species and included the greatest proportion of the vegetation in the

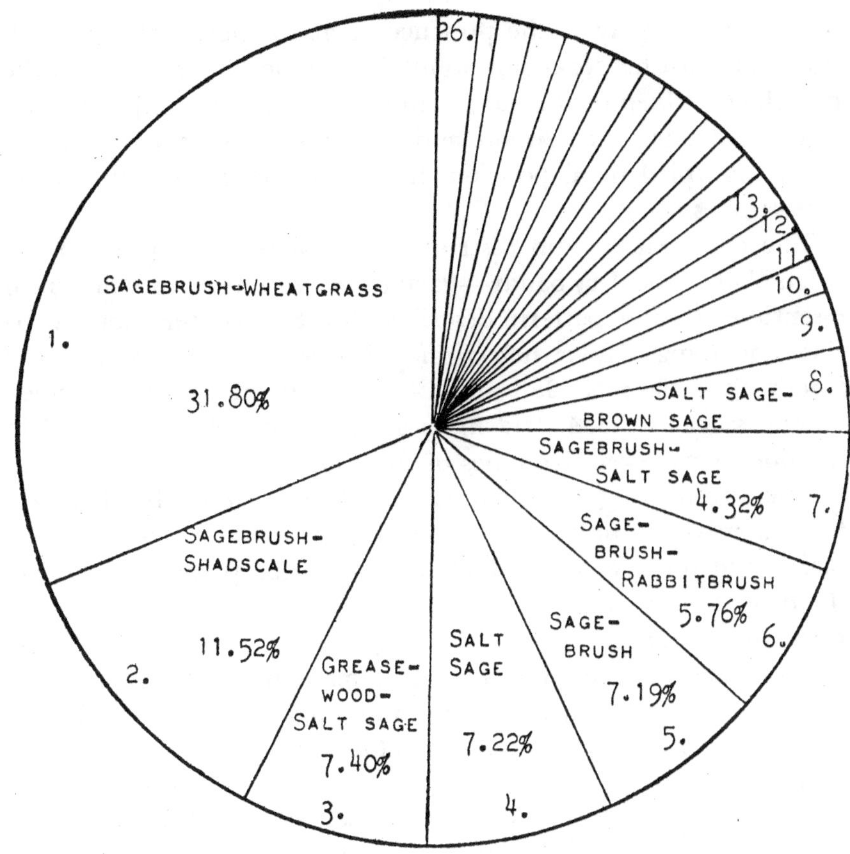

Dominant	Sub-dominant	Dominant	Sub-dominant
1. Artemisia tridentata-Agropyron sp.		14. A. nuttalli-Sarcobatus	
2. A. tridentata-Atriplex confertifolia		15. Eurotia-A. tridentata	
3. Sarcobatus-Atriplex nuttallii		16. A. nuttallii-Eurotia	
4. A. nuttalli		17. A. tridentata-Oryzopsis hymenoides	
5. A. tridentata		18. A. pedatifida-Agropyron sp.	
6. A. tridentata-Chrysothamnus sp.		19. A. confertifolia	
7. A. tridentata-A. nuttallii		20. Chrysothamnus-Distichlis	
8. A. nuttallii-Artemisia pedatifida		21. A. confertifolia-A. nuttallii	
9. A. tridentata-Eurotia lanata		22. A. tridentata-A. pedatifida	
10. A. nuttallii-A. tridentata		23. Sarcobatus-Atriplex pabularis	
11. Sarcobatus vermiculatus		24. A. nuttalli-Agropyron sp.	
12. Chrysothamnus-A. tridentata		25. A. pedatifida-A. nuttallii	
13. A. nuttallii-A. confertifolia		26. All other types	

FIGURE 5. PROPORTION OF SWEETWATER COUNTY
OCCUPIED BY VARIOUS VEGETATIVE TYPES
(After Vass and Lang, 1938)

county, due largely to the presence of many species of sagebrush and rabbitbrush. Next important in number of species was the goosefoot (*Chenopodiaceae*) family, represented by the salt sages, shadscales, greasewood, and winterfat. The grass (*Poaceae*) family occupied third position in the number of relatively common species.

The composition and distribution of vegetative types on the Eden Valley and Dry Sandy—Pacific Creek study areas conformed essentially to the pattern described for Sweetwater County by Vass and Lang. Nevertheless, they did not recognize any of the sagebrush variads of Hall and Clements in the sagebrush types. Several sages of the *A. tridentata* group were well-represented in the dominant vegetative types of the county. On the study areas the big sagebrush (*A. tridentata*) and black sagebrush (*A. nova*) formed mixed stands, the former being dominant on outwash plains and on stream flood plains, the latter predominating on the drier slopes and ridges.

Plant Succession on Sagebrush Ranges

Under the natural conditions of climate and soil now existing on semiarid ranges in the intermountain region, the sagebrush-grass type will maintain itself indefinitely as the climax vegetation.

As a result of improper land practices, numerous areas in this region which originally supported pure grass types have reverted to the sagebrush climax. One such area has been desribed by Cottam and Stewart (1940) in southwestern Utah. The Mountain Meadows Valley in Utah was a relatively small tract of grassland thrust in the midst of an extensive sagebrush climax. It originally had good water and meadows filled with luxuriant grasses. Following settlement in 1862, the area was heavily grazed by cattle and sheep. Depletion of the plant cover permitted torrential rains in 1884 to wash deep gullies which ultimately drained most of the meadow. The grassy vegetation has now disappeared and has been replaced largely by sagebrush, rabbitbrush, and juniper. Cottam and Stewart concluded that the invasion by sagebrush and juniper has retarded the recovery of range grasses.

Additional studies on plant succession in sagebrush types have been conducted by the Forest Service on the Beaverhead National Forest in southwestern Montana (Lommasson, 1948). Heavy grazing by buffalo, prior to 1882, reduced grassland meadows in the mountains to areas of bare soil pocked with buffalo wallows. Following the extermination of the buffalo, little grazing use was made of this range. By 1914, sagebrush had become well-established in favorable locations. Continued studies on this area by the Forest Service indicated that the sagebrush would maintain itself indefinitely under natural conditions. As a result of this investigation the Forest Service predicted inevitable reduction of grazing capacities in mountain grasslands being threatened by sagebrush invasion and dominance. In order to forestall a large-scale sagebrush eradication and control program in the future, the Service has sounded a clarion call for immediate action to eradicate sagebrush in the incipient stages of invasion of mountain grassland meadows.

On the semiarid foothills of the upper Missouri River Valley in Montana, clean burns of mature sagebrush have revegetated to a thrifty stand of young plants, five years after fire had completely removed the old stand (Lommasson, 1947). The experimental burn was made by the Forest Service in an attempt to improve grazing conditions for sheep. The thriftiness and evenness of the stand five years after burning indicated future dominance by sagebrush, which in the opinion of the Forest Service would again render the land unsuitable for grazing.

It is to be expected that sagebrush will invade grassland areas such as described above, when the original soil condition has been altered by overgrazing, erosion, and drainage. Such invasion by sagebrush would not necessarily prevent the eventual reestablishment of a subdominant stand of native grasses under proper range management and restoration of the original soil conditions.

Elimination of the sagebrush climax on most areas of the Green River Basin has been followed with initial invasion by Russian thistle (*Salsola pestifer* A. Nels.), followed by forbs, grasses, and rabbitbrush. It often required several decades before sagebrush again regained its dominance on disturbed areas. An airstrip constructed in Eden Valley in 1934 has revegetated primarily to rabbitbrush and bunch grasses over a period of sixteen years. Small sagebrush plants have just recently appeared on this area. Similar patterns of plant succession have occurred on Eden Valley homesteads which were abandoned. In mountainous areas with higher annual precipitation, sagebrush invaded denuded plots at a more rapid rate.

Food and Shelter Relationships Between Sage Grouse and Plants

The close relationship between sagebrush distribution and sage grouse abundance has been more than just casual occurrence. As a primitive wilderness area, the sagebrush-grass formation has been characterized by distinct unity in the patterns of vegetation, climate, soil, and physiography. The dominant plant species have evolved in accordance with an established pattern of composition and arrangement over geologic periods of time. Sage grouse developed under all of these influences and, therefore, represent a true climax inhabitant of a climax vegetative type.

The nature of the environment has placed heavy demands upon the physiological system of the sage grouse. Long winters, dry summers, high elevations, heavy snowfall, and ground blizzards all have provided critical hazards in the search for food, shelter, and water. Chemical analyses of sagebrush have disclosed abnormally high concentrations of fats in the leaves. Since sagebrush is an evergreen plant, its leaves are available at all seasons as a source of food and shelter. There is little variation in the chemical composition of sagebrush leaves throughout the season. On ideal sites, the shrub varies its growth form from plants a few inches tall to several feet in height. Consequently, it is seldom unavailable to sage grouse on a major drainage. Nearly all other plants of the sagebrush prairie possess deciduous leaves. Studies on the food and shelter requirements of sage grouse have revealed that they are solely dependent upon sagebrush during the winter months. At other seasons sagebrush furnishes the bulk of their food and shelter demands.

The extensiveness and unity of sagebrush types have contributed to the development of an extremely specialized type of feeding habit in sage grouse. In contrast with most other species of upland game birds, the sage grouse has a digestive system which is not adapted for a wide variety of plant foods, especially the hard seeds and grains.

Development of Vegetation and Sage Grouse Utilization

The form of the individual sagebrush plant varies from a low, prostrate shrub a few inches in height on dry, rocky sites to a bush of tree-like proportions on moist, sandy soils along water courses. In general, the density and height of sagebrush increases from the semiarid desert ranges toward the mountain foothills. The various growth forms of the plant furnish essential increments of sage grouse habitat.

For purposes of courtship, strutting, and mating, sage grouse habitually selected areas supporting either a low, sparse covering of sagebrush or else an area denuded of vegetation.

Nesting areas were normally located in sagebrush cover varying in height from 1 to 2 feet and covering considerably less than 50 per cent of the ground surface or area. Eighty-two per cent of some 300-odd nests studied over a three-year period were located in cover types varying from 10 to 20 inches high. The average height of all nesting cover was 14 inches. Loafing and roosting grounds were invariably selected in areas supporting the heaviest and densest sagebrush cover. Optimum loafing and roosting cover was found along stream bottoms, ravines, and draws. Feeding areas were situated in sagebrush stands of intermediate height and density. Sage grouse characteristically fed while standing on the ground level. Birds have never been observed to climb or fly into the branches of sagebrush for purposes of feeding. Snow conditions naturally altered the ground feeding patterns. In the winter, birds may be forced to feed on the exposed tips of sagebrush branches or compelled to pick the leaves from small prostrate bushes on windswept ridges. Griner (op. cit.) recorded instances in Utah of sage grouse digging holes in the snow to reach the tips of sagebrush leaves. Observations in Wyoming have revealed the importance of antelope in making snow-covered sagebrush available to sage chickens as a result of digging for their own food supply.

To some extent, the utilization of different types of sagebrush growth forms may be altered by conditions of site. For example, in the Jackson Hole region a remnant sage grouse population continues to thrive in uniformly dense and heavy sagebrush flats bordering the Snake River flood plain. This growth condition is characteristic of islands of sagebrush in isolated intermountain regions. Semidesert and foothill sagebrush ranges seldom display such uniformity of growth patterns.

Animal Environment

The Sage Grouse and its Animal Associates

The upper Green River Basin has long been one of North America's most fabulous wildlife areas. Featuring several major wilderness areas from the semiarid desert to the alpine tundra, it probably supports a larger variety and greater numbers of game and other wildlife species than any area of comparable size on the continent. In close proximity are located the arctic-alpine tundra, the spruce-fir forest, the lodgepole pine forest, the sage-brush-grass prairie, and the salt-shrub desert. The seasonal changes in climate force a host of animals to move from one large plant formation to another, especially species inhabiting the tundra and forest, some of which annually move onto the desert during the winter season. In addition to providing for its resident animals, the sagebrush desert furnishes range for this seasonal influx of migratory birds and mammals from life zones lying at higher elevations.

Big Game—It has been significant that herds of antelope were closely associated with flocks of sage grouse at all seasons of the year in the Green River Basin. Antelope, like sage grouse, were dependent upon sagebrush and associated plants for their seasonal habitat requirements. Both of these animals exhibited seasonal movements from summer to winter ranges in essentially the same areas and at the same periods. Their sharing of winter ranges was particularly evident during the biennial statewide aerial antelope censuses which have been conducted in Wyoming during January and February.

Antelope populations in Wyoming have had a history of abundance and decline similar to the fluctuations of sage grouse. It has been estimated that at the peak of pronghorn abundance in the middle 1800's, the antelope herd along the Big Sandy Creek in the Green River Basin numbered at least 30,000 animals. By the early 1900's, antelope numbers in all of Wyoming totaled less than 5,000 head, with the largest herd being located along the Green River. From this low point, antelope populations in the state have staged a steady comeback. The number of antelope in Wyoming now exceeds 92,000, in spite of severe winter storms in 1948-49 and notwithstanding a total harvest of 22,792 animals in 1950 and 39,315 in 1951. According to the 1952 aerial census, approximately 7,000 antelope were located on the upper Green River drainage in Wyoming. The antelope harvest in this area in 1950 was about 1,850 animals, and in 1951, 2,650 antelope were legally killed in the same sector.

Historical accounts have verified the occurrence of annual migrations of elk and deer onto the sagebrush desert from summer ranges in the mountains of the Green River Basin. Ranches, fences, and civilization in the foothills region now largely have cut off these migrational paths to the desert. However, efforts by the Game Commission to restore some of these earlier migration routes have been successful, and at the present time several hundred elk annually winter on the sagebrush desert (Allred, 1950). During severe winters hundreds of mule deer move onto winter ranges on the desert. In 1949-50, large herds of mule deer and antelope shared winter ranges in the Eden Valley area. The deer had been forced out of the foothill ranges from 50 to 75 miles distant, as a result of food shortages caused by deep, hard-crusted snow. Small numbers of elk have annually wintered near the Eden Valley irrigation district and on several occasions have inflicted damage to haystacks and fences. It is a mild winter, indeed, when a few moose do not winter in the dense willow thickets along desert water courses near Eden Valley. In recent years, moose have been spreading southward from the Wind River Range. They are occasionally observed in the Red Desert region, and are even rarely found in northern Colorado.

Waterfowl—Contrary to popular belief, the semiarid regions of Wyoming are important breeding and resting grounds for myriads of waterfowl. Although the size of individual waterfowl habitat in Wyoming is relatively small, thousands of desert water holes, springs, marshy meadows, and miles of streams furnish breeding areas which are tremendously important in the continental waterfowl picture. This is especially true, since one of the main objectives in waterfowl management is the decentralization of breeding and resting areas. In some cases the original habitat has been supplemented and improved by water development programs and irrigation structures associated with reclamation projects, large and small. The various reclamation projects have been functioning essentially as migratory havens, since the large water impoundments furnish safe resting areas and the adjoining agricultural areas supply large quantities of grain for food. Most of the ducks and geese in Wyoming retire to local agricultural areas following the nesting and rearing seasons. In the past, little has been known of the state's contribution to waterfowl production, or of seasonal movements of waterfowl through the state. It is hoped that recently inaugurated waterfowl investigations will demonstrate Wyoming's important role in waterfowl production and management.

Observations in Eden Valley during the years from 1949 to 1951 have revealed that several thousand migrant ducks spent considerable time in the valley during the spring and fall. Sixteen species of surface-feeding and diving ducks, as well as three species of geese, have been recorded in Eden Valley during this study. Substantial numbers of the surface feeders have remained to nest in the Green River Basin in the habitat previously described. Mallards, pintails, green-winged teal, and gadwalls have been the major resident nesting species. These ducks commonly nest in sagebrush areas adjacent to some sort of open water. Mallard and pintail nests were found on several occasions on sage grouse nesting plots 1 to 2 miles from open water. Springs and stock water impoundments located on the sagebrush desert invariably produced a brood or two of mallards, pintails, or teal every season.

The Green River system is an important nesting area for at least several hundred pairs of Canada geese. The Craigheads (1949) have demonstrated the value of mountain river systems as nesting areas for Canada geese in their studies along the Snake River in northwestern Wyoming. Several pairs of geese have nested annually along the streams and reservoirs in Eden Valley. One nest in particular has been located for three consecutive years in a former ferruginous rough-legged hawk nest on a cliff overlooking Big Sandy Creek.

Upland Game Birds—In many cases, reclamation projects have almost exterminated sage grouse populations due to the complete elimination of sagebrush habitat and replacement with cultivated crops. On such areas exotic game birds have largely replaced native sage grouse populations. Pheasants have been widely introduced and are well-established in most of Wyoming's agricultural areas below 5,000 feet elevation. Hungarian partridges are locally abundant in agricultural areas on the drainages of the Powder and Tongue rivers in northeastern Wyoming. The fringes of most reclamation areas in sagebrush country provide marginal habitat for remnant pockets of sage grouse which are more or less in competition with the introduced game birds. Chukars are being planted over extensive areas of the state and frequently on sagebrush prairies supporting sage chicken populations. Little study has been devoted to an evaluation of game bird planting programs in Wyoming, particularly with attempts to establish exotic forms within the ranges of native upland game birds. In addition to the sage grouse, the native sharp-tailed grouse is found abundantly on semiprairie regions of northeastern Wyoming. It occurs in close association with sage grouse along the periphery of their native ranges. Mourning doves occur commonly throughout the semiarid prairie, as well as in agricultural areas of Wyoming. Dove nests were found on the ground in sagebrush types in Eden Valley.

Within the Green River Basin there have been only scattered attempts to establish exotic game birds, mainly chukar partridge, Hungarian partridge, and pheasants. Pheasants have become established on a small scale within the agricultural portions of the Eden Valley irrigation district. Based upon crowing counts taken in 1949 and 1950, the pheasant population in Eden Valley approximated fifteen birds per square mile of irrigated land. The other species have disappeared shortly after planting and have not become established in any areas of the Green River Basin.

Intensive studies of the relationships between pheasants and sage grouse have been conducted in Eden Valley. There seemed to be little direct competition between the two species either for food, cover, or breeding territory. They did, however, indirectly compete for habitat in the sense of space. Any area supporting pheasants, or other introduced game birds, can no longer be considered as optimum sage grouse habitat, even though the two species may intermix during certain seasons of the year. This sort of pattern occurs when sage grouse move into agricultural fields for purposes of feeding and again when pheasants take refuge during the hunting season in sagebrush areas adjoining extensive farming lands.

At the point where sagebrush adjoins wooded forest types, sage grouse ordinarily are found in close proximity to ruffed grouse and blue grouse. This situation is encountered in the mountain foothills region and along timbered river bottoms of streams such as the Green and Snake rivers.

Overzealous sportsmen and game administrators have introduced pheasants and chukar and Hungarian partridges at many points within the natural range of the sage grouse and are continually searching remote regions of the earth for additional game birds to establish on the prairies occupied by native grouse. From the standpoint of preferred habitats and direct competition for food, shelter, and water, there may be relatively little conflict between sage grouse and a host of exotics. However, there are other hazards and important considerations incident to the release of exotic birds on native game bird ranges, such as hybridization and resultant deterioration of native forms, reduction of the total capacity of the range to support native birds due to infiltration of exotics from the habitat originally intended for them, and a gradual lack of interest on the part of the public in

Sex may be distinguished by the appearance of the throat and breast region

Centrocercus urophasianus—"a spiny-tailed phasant"

Sagebrush (*Artemisia tridentata* and closely related types) provides all of the food and shelter requirements of sage grouse during the winter months (November-March)

Mountain deserts constitute the home of sage grouse. Teton Mountains in the background

Sage grouse habitat in Eden Valley—sagebrush-grass type

Rabbitbrush (*Chrysothamnus*) invading an abandoned homestead site in Eden Valley. Wind River Mountains in the distance

Big Sandy Creek meandering through Eden Valley

Alkali flat and associated salt-desert shrub type—salt sage (*Atriplex nuttallii*) with greasewood (*Sarco-

High-speed roads and luxuriant roadside vegetation produce a heavy grouse highway mortality. Dry Sandy—Pacific Creek area

Characteristic sign of the sage chicken poacher (Pacific Springs, August 1, 1949). *Photo by Warren W. Chase*

Sage grouse tracks in deep, powdery snow

Sage grouse tracks on light snow

Typical grouse winter roosting site and tell-tale pile of intestinal droppings and accompanying black caecal dropping

Comparison of sage grouse eggs (spotted) with pheasant eggs

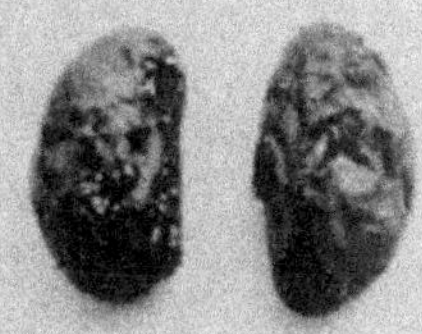

Testes of an old cock as compared with those of a
yearling cock on March 18

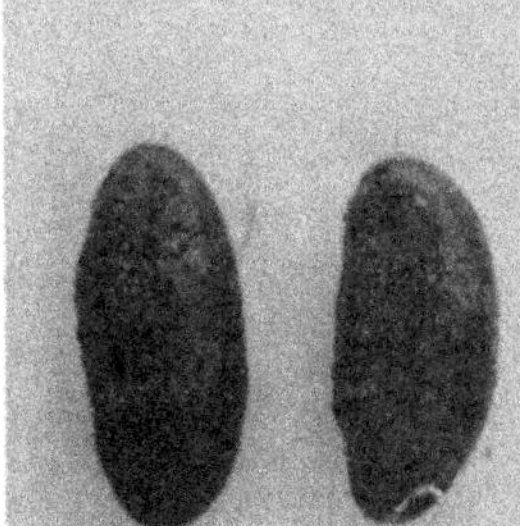

Testes of an old cock as compared with those of a
yearling cock (now sexually mature) on April 23

Wings of an adult hen and a yearling hen in April, showing the character of the two outer primaries

the native species as the exotics become popular. Another consideration is that the introduction of exotic game birds onto ranges featured by annual hunting seasons on native game birds subjects the protected species to an undesirable illegal hunting pressure. During the 1950 and 1951 sage grouse hunting seasons in Eden Valley there was a substantial illegal kill of pheasants in the agricultural areas, although much of this kill was unintentional as evidenced by the fact that several hunters stopped at permanently established game checking stations with a mixed bag of pheasants and sage grouse.

Careful evaluation of the status of native upland game birds should be an essential part of prospective planting programs involving exotic game birds. Valuable species are now being displaced from their native ranges at an alarming rate, without speeding the process through media of poorly considered planting programs at tremendously high costs to the sportsmen.

Birds of Prey—Due to wide variations in topography and vegetative patterns which exist in the western mountain foothills region, significantly more species and higher densities of both migrant and resident birds of prey are found here than on areas of equal size in eastern United States. Studies by the Craigheads (1950) on comparative breeding densities of hawks and owls in southern Michigan and northwestern Wyoming revealed that the Michigan township of mixed agricultural and woodlot composition supported a hawk and owl population of one and one-half breeding birds per square mile; whereas an area of similar size in the primitive Jackson Hole region of northwestern Wyoming supported a raptor density of approximately three breeding birds per square mile.

Within the Eden Valley study area, twenty species of hawks and owls were recorded either as migrants or summer residents during the course of this investigation. Ten species were observed to breed on the two study areas. The most common breeding species were the Swainson's hawk, ferruginous rough-legged hawk, western red-tailed hawk, marsh hawk, prairie falcon, great horned owl, and burrowing owl. With the exception of the burrowing owl, the main nesting areas of these birds of prey were located in the immediate vicinity of the stream courses

which pass through the Valley. The presence of scattered cotton-
wood trees, willows, and occasional overhanging cliffs furnished
ideal nesting sites for birds of prey. On areas adjacent to Eden
Valley, the eroded escarpments of buttes, ledges, and pinnacles
were preferred nesting sites for golden eagles, ferruginous rough-
legged hawks, and red-tailed hawks. Swainson's hawks utilized
cottonwoods and willows primarily as nesting sites, although
nests were occasionally placed in the tops of large sagebrush
bushes. Raptor nesting densities along stream courses such as
Big Sandy Creek approximated four breeding birds or two nests
per linear mile of stream. Nesting densities were reduced considerably over the remainder of the Eden Valley study area, due
to the scarcity of potential nesting sites.

In the absence of a food habits study involving all raptors on
the sage grouse study areas, it has been somewhat difficult to
estimate the extent of predator pressure upon sage grouse populations. Sage grouse nesting studies revealed that hawks and owls
were unimportant as a source of nest destruction. However, magpies were responsible for considerable nest destruction in the
vicinity of ranching areas. Since they were seldom encountered
outside of agricultural areas, magpies did not ordinarily share a
prominent role in nest destruction in unsettled areas. As additional regions are developed for settlement, the magpie may be
expected to assume increasing importance as a factor in sage
grouse nest destruction. There was little question that predation
by certain raptors upon juvenile sage grouse was of normal occurrence. During the nesting period certain broad winged
hawks, e.g., red-tailed and ferruginous rough-legged, utilized
young sage grouse as a source of food for their own offspring.
The remains of young sage grouse were commonly found at nest
sites of ferruginous rough-legged hawks. Sage grouse remains
were seldom encountered at great horned owl and prairie falcon
nests observed in Eden Valley, though both of these species are
capable of killing adult birds. Other avian predators encountered on the area would not ordinarily attack adult sage grouse.
No golden eagle nests were located on the study areas. The ferruginous rough-legged hawk was the most common raptor on
sage grouse summer range. Although this hawk was widely dis-

tributed, the birds were not concentrated in sufficient numbers to seriously affect the survival of an entire juvenile population of sage grouse on any district. On the study areas, ferruginous rough-legged hawks nested exclusively on cliffs, buttes, pinnacles, and ledges. In few areas would breeding densities of the hawk have exceeded one nest for every 10 square miles. The major sources of food for juvenile and adult birds of prey consisted mostly of small rodents, rabbits, small birds, and insects.

Golden eagles were distinctively associated with winter flocks of sage grouse in the Green River Basin. Car censuses through areas of sage grouse concentrations during the 1949-1950 winter season revealed an average of one golden eagle for every 6 miles of traversed distance. In December, 1950, a check of 14 miles of cross-country telephone line disclosed sixteen golden eagles perched on the poles in that distance. It was a common occurrence to observe from fifteen to twenty-five golden eagles during a single afternoon's field activities over winter grouse range by car. After being disturbed from their hunting perches, golden eagles were frequently observed to flush flocks of sage grouse. At the appearance of golden eagles, sage grouse were either frightened into wild flight or else froze in a crouching position. On no occasion did this worker observe eagles to kill sage grouse, although many persons have undeniably witnessed this act. On several occasions golden eagles were seen to stoop on sage grouse as the latter alighted on the ground. Eagles were flushed several times from freshly killed carcasses of adult birds. During the early strutting season eagles frequently disturbed cocks on the strutting grounds, causing the entire assemblage to fly. On most strutting areas a few sage grouse were victims of eagle predation, as evidenced by the carcasses and feathers found scattered over the grounds. The abundance of eagles in the vicinity of winter flocks of sage grouse, as well as the wild flushing of the birds whenever eagles flew near, indicated that sage grouse undoubtedly were an important prey species for golden eagles during winter months. Most golden eagles encountered in this area during the winter season migrated to the higher mountains or to regions northward of the Green River Basin for nesting purposes. With the exception of some nonbreeding birds, few golden eagles were located in areas heavily utilized by sage grouse during the breeding season.

Avian predator control campaigns in Wyoming have not been concentrated on hawks and owls. In fact, the game laws of the state extend full protection to almost all hawks and owls, notably excepting the golden eagle, great horned owl, and the accipiters (blue darter hawks). It has been noteworthy that, in spite of the large numbers of raptors which have been in close association with sage grouse populations, the latter have flourished and increased to levels permitting liberal open seasons throughout the Green River Basin. Predator control campaigns certainly played no role in the maintenance of high game populations in primitive times, and there is little data to indicate that present day control programs merit the full credit of success heaped upon them.

Mammalian Predators—On the basis of sight records and sign encountered in the field, coyotes were considered to be uncommon over the sage grouse study areas during 1948-1950. Only two coyotes were seen in the course of field observations during the fall and early winter of 1949-1950. Both these animals were observed within 100 feet of perched golden eagles. On other occasions this same association or symbiotic relationship between eagles and coyotes has been noted by this author and also by personnel of the U. S. Fish and Wildlife Service. In the fall and winter seasons of 1950-51, coyotes again seemed to be uncommon on the sage grouse study areas, since only two were seen. There was very little predation by coyotes upon sage grouse nests. In view of the low number of coyote sight and track records in the vicinity of sage grouse flocks, it was not considered that they were preying upon the grouse to any great extent. Information supplied by a local coyote den hunter revealed that in 1951 three pairs of coyotes had denned on the Dry Sandy—Pacific Creek area and four pairs had denned in Eden Valley. The apparent absence of coyotes over parts of the sage grouse study areas was not typical of other areas in the Basin, as evidenced by the catch records of the Branch of Predator and Rodent Control of the U. S. Fish and Wildlife Service (Table 48), and by experiences of livestock operators. Coyotes have continued to exert a constant predation pressure upon domestic sheep herds, and in spite of intensive control many operators sustained losses from coyotes, especially during the lambing season. There was, however, an acknowledged tendency upon the part of sheep herders to attribute losses from all causes to predation.

On sage grouse nesting plots in the Dry Sandy—Pacific Creek area, badgers and ground squirrels were major sources of nest destruction. Upon first analysis, it might be assumed that they were seriously limiting the favorable increase of the birds in many areas of the Basin. However, nesting and strutting ground censuses revealed only minor changes in spring breeding populations on any of the study tracts during 1949 and 1950. The sage grouse population seemed to be in balance with its environment, even though nest destruction by mammalian predators accounted for a relatively high percentage of all nests.

From the management viewpoint it might be desirable to reduce the incidence of nest destruction by badgers, provided that this could be accomplished on a normal harvest basis by private fur trappers and not through an organized predatory animal control program. It must be realized, however, that badgers act as a natural check upon the increase of ground squirrels which also prey heavily upon sage grouse nests.

Other Birds and Mammals—Check lists of all birds and mammals observed by this worker in sagebrush types of the Green River Basin, including those species known to occur but not recorded, appear in the Appendix of this report (Tables 51 and 52).

Development of the Sagebrush-Sage Grouse Community

The considerable number of herbaceous species of *Artemisia* common to Eurasia and North America has led to the belief on the part of some botanists that it was originally a boreal genus which has extended far southward in recent geological time. This movement brought it into regions characterized by drier climates, and subsequently led to the evolution of a new group of plants which assumed a shrubby habit in response to arid climates. According to Hall and Clements (op. cit.) it was probable that *Artemisia tridentata* was one of the first shrubby species that developed in response to aridity in the southwestern United States and Mexico. During the dry phase following the Pleistocene era they believe that it moved northward, occupying the southern half of the Great Basin as the great dominant plant and extending northward into Idaho, Wyoming, Oregon, and Washington to form a low savannah in the bunch grass and mixed-prairie types. With the advent of the historical period, grazing so reduced the native grasses that much of the savannah was converted into a pure sagebrush community.

The sagebrush climax has been regarded by Weaver and Clements (1938) as a desert scrub which has migrated widely since its early development in the southwest. It differs from the desert scrub, mesquite, and chaparral climaxes, in that practically all of the dominant plants are shrubby adaptations of predominantly herbaceous families such as the *Compositae* and the *Chenopodiaceae*. Due to the nature of its bushy growth form and herbaceous branches, sagebrush has adapted itself to survival under a wide variety of climatic conditions, thriving as the climax vegetative type in regions of 5 to 10 inches of precipitation and extending into areas of 10 to 20 inches where soil conditions or elimination of native grasses by intensive land use favored its growth and development.

The sagebrush climax is usually confined to intermontane basins, high plateaus, and glacial outwash plains. It is in these situations that the sage grouse has attained its highest populations and greatest development. The Green River Basin is such an area. It has long been an important habitat for sage grouse. The pronghorn antelope has been another distinctive climax species; possibly it developed simultaneously with the sage grouse in the sagebrush-grass community, since both animals utilize the same habitat, and in the past have flourished and declined in patterns of unusual similarity.

Water and Sage Grouse Abundance

In primitive times, water was undoubtedly a key factor governing the distribution and relative abundance of sage grouse populations. It is equally important now, although it often has been artificially created as a feature of agricultural programs, grazing developments, and irrigation systems. Whereas these latter programs have upon occasion improved sage grouse habitat locally, they more frequently have drastically reduced sage grouse numbers as a result of the vast acreage of sagebrush destroyed under the newer land-use programs.

Although water's dominant influence upon sage grouse has been reflected in the development of the sagebrush climax, weather and climate are locally responsible for extended grouse movements resulting in a pronounced seasonal distribution. The foothills, so attractive to sage grouse in the spring and summer, become untenable in the winter due to the snow depths over the food supply. At this season the normally arid desert regions are blanketed in lighter snowfalls than are the foothills, and provide accessible moisture and exposed sagebrush for food. During the summer months sage grouse are normally limited in their desert distribution to the immediate vicinities of stream courses, isolated desert springs, and water holes.

On the Dry Sandy—Pacific Creek study tract there was evidence to indicate that hens retired to the foothills region and more abundant water supplies during unusually dry seasons. Plots located midway between the mountains and Eden Valley produced significantly lower nesting densities in 1949 than were encountered on plots nearer the foothills and on plots in Eden Valley. Streams in this central area normally dried up in late spring or early summer due to the lack of precipitation on their watersheds which lay in the foothills rather than in the high mountains. In 1950 snows lay deep over the foothills and extended well onto the desert, reaching the Eden Valley area. There were indications that hens selected more widely dispersed areas for nesting in 1950 than was the case in 1949. The plots described above produced six nests in 1949 as compared with twelve in 1950. These plots totaled 200 acres and were located in a region characterized by only small amounts of precipitation in normal spring and summer seasons. In 1950 the late spring season and abnormally heavy snow cover created large amounts of open water. This was responsible for an abundance of subdominant grasses and forbs throughout large areas of the desert.

On the basis of the behavior displayed above, it was felt that

a series of drought years would likely telescope sage grouse populations into areas less likely to exhibit the effects of continued
drought as expressed in the vegetation and available surface
water. This phenomenon would be expected to occur in almost
all of the sage grouse habitat with exception of the areas lying
adjacent to the mountains. It would, in effect, not only result in
a restriction of sage grouse distribution, but would cause a
noticeable decrease in numbers on submarginal ranges. Such a
phenomenon occurred during the mid-1930's throughout much
of the birds' range. Although accurate population figures for this
period are not available, we do know the general trend of numbers from the reports of game wardens and sportsmen. It was
during this period that Wyoming and other states either closed
the season on sage grouse or else instituted greatly curtailed
shooting seasons.

Stream runoff figures and weather records in the Eden Valley
area substantiate the occurrence of below normal precipitation
for several years in the early 1930's (Tables 4 and 5).

It has been generally recognized that the amount of forage
and vegetative composition on western range lands has fluctuated
widely in years of drought and high precipitation. In regard to

TABLE 5

*Annual Precipitation in Eden Valley—1930-1940**

Year	Precipitation
1930	9.58 in.
1931	4.87
1932	8.00
1933	6.60
1934	5.69
1935	6.21
1936	13.70
1937	11.40
1938	12.82
1939	4.87
1940	8.17

Mean — 7.24 in.

*Data from U. S. Weather Bureau

the changes in range vegetation caused by the drought years of the mid-thirties, Savage (1937) states that the basal plant cover declined from an estimated average of 90 per cent or more on certain areas before the drought to an average of 25 per cent or less on most of the ungrazed areas after the drought. This estimated reduction of about 65 per cent on certain areas was due wholly to the effects of climatic extremes. The difference between the plant cover on the grazed and ungrazed areas after the drought usually amounted to about 10 per cent, which may be considered the additional injury caused by grazing alone.

In a similar manner, Holscher (1944) reported that on the northern Great Plains total vegetation in the spring of 1937 was only 10 per cent of what it had been in 1933, before the drought years. By 1943, the vegetation on the experimental range had recovered 92 per cent of the area it had lost as a result of the drought.

Based upon an eight-year study to determine the relationship between precipitation and density of native vegetation in Wyoming, Lang (1945) found a striking relationship between the precipitation during any twelve-month period beginning September 15 and the condition of the vegetation in the following growing season. The density of the perennial grasses decreased noticeably during drought years. There was a general increase in the densities of annual grasses and forbs, whereas the shrub portion of the vegetative cover tended to remain relatively constant. Since the period of study covered some extremely dry years, as well as some relatively wet years, it has shown that density decreases of 50 per cent or more may be expected from wet to dry years. During the course of the experiments, one area was afforded complete protection from grazing for three years. Although the total density of vegetation did not materially change, there was a slight increase in grass and shrub density and a corresponding decrease in forb density on the protected area. Although the total density had not been affected by protection, the composition of the vegetative cover was being slowly altered.

All these studies seem to agree that large reductions in range forage occurred during drought years, and, furthermore, that many valuable forage plants were killed. However, with the renewal of average precipitation, as well as proper management during the dry years, it can be expected that the range will quickly recover its productiveness.

It is unlikely that the effects of either drought or overgrazing would reduce the density of sagebrush to a point where it would no longer be available as a source of food for sage grouse. On the other hand, the reduction and elimination of perennial grasses and forbs will impose serious restrictions upon the normal feeding habits of females and young birds, particularly during the early stages of the latter's development. At this time they are highly dependent upon an insect diet. In order to obtain this type of food, broods customarily disperse from the individual nest site and concentrate in the vicinity of native grass and sedge meadows. During the spring and summer months juvenile and adult sage grouse depart from a pure sagebrush diet and utilize a wide variety of herbaceous material, as well as insect food.

Sage grouse populations in the mountainous regions fluctuate less widely during the drought years than do birds on the deserts. Following years of subnormal rainfall, the birds disperse from the mountains and provide a source of brood stock for replenishment of the species in the more arid regions of their range.

Diseases and Other Hazards Affecting Sage Grouse

Nature of Disease in Sage Grouse

There has developed a widespread belief in Wyoming that the element of disease has been largely responsible for heavy annual losses and a rapid decline in sage grouse numbers within the state. Similar ideas have developed among sportsmen throughout the country concerning game bird diseases. To the average person, the presence of parasites and disease producing organisms is synonymous with "disease" itself, whereas in the majority of instances the presence of some parasites is of normal occurrence in all wildlife species and normally does not cause any significant alteration in structure or function of the host species. The parasites and disease organisms affecting sage grouse have conformed essentially to this pattern of occurrence, although in the case of coccidiosis small epizootic outbreaks apparently have developed over widely separated areas in Wyoming and adjacent states in recent years. However, there is insufficient data on either the incidence of the disease or on sage grouse populations to indicate that the disease of coccidiosis was in any way related to the major decline in sage grouse nunmbers over any extensive area of the birds' range.

With relation to diseases of sage grouse, the work of this present investigation has primarily involved co-operation with the Disease Research Laboratory maintained jointly by the Wyoming Game and Fish Commission and the University of Wyoming at Laramie. In all cases, parasitized and diseased sage grouse encountered in the field have been forwarded to the Disease Laboratory for examination and a report of pathogenicity. Personnel of the laboratory have undertaken specific projects relating to parasitism and diseases of sage grouse as entities separate from this study. Primary consideration has been given to coccidiosis and its relation to sage grouse populations.

Parasites of Sage Grouse

Felix Simon (1940) has catalogued the parasites of the sage grouse (Table 6), and included in his report essential elements of the original descriptions of the parasites. The parasites were taken from about a hundred birds collected throughout Wyoming. With the exception of the ticks reported from Montana, and one record of *Heterakis gallinae,* correspondence by Simon with parasitologists throughout the range of the sage grouse revealed no species of parasite which had not been taken from grouse in Wyoming. Simon concluded his study with the following summary of the pathogenicity of the important species of parasites encountered in sage grouse:

> Great numbers of *Tritrichomonas* sp. were found in every sage grouse whose caecal contents were examined microscopically. It seemed to be harmless, although it occurred in vast numbers and the incidence of the infection was very high.
>
> *Eimeria* ssp. (coccidia) have been known to decimate sage grouse populations in several localities.
>
> Although there was a high incidence of *Raillietina centrocerci* and many of these large tapeworms were often present in a single host, no serious ill effects were noted.
>
> *Cheilospirura centrocerci* produced necrosis in the horny tissue of the gizzard as well as in the underlying muscles.

Also, a slight hemorrhage sometimes accompanied the necrotic condition, but in no case did the general health of the bird seem seriously affected.

Habronema urophasiana, whenever found, was associated with *Eimeria centrocerci* and, therefore, its pathogencity could not be judged.

Haemaphysalis cinnabarina (tick) was shown by Parker et al. (1932) to be the proabable vector of tularemia, and is, therefore, a great menace. Although no records of ticks from sage grouse in Wyoming were reported, sage grouse in Montana were heavily infested with *Haemaphysalis cinnabarina.* Since these ticks were found in such numbers, they must have been a considerable nuisance to sage grouse.

With the exception of the protozoan *Tritrichomonas,* the incidence of all other parasites was definitely higher in the young birds than in the adults; also, the young were more heavily infested.

No injurious effects were attributable to the other parasites encountered in sage grouse.

Honess (1947) reviewed the literature relating to coccidiosis in sage grouse in Wyoming, and presented some new information. In September, 1932, Scott and Honess investigated an epidemic disease of sage chickens in Fremont County, Wyoming. The center of the outbreak was a ranch located on the Sweetwater River about 28 miles northwest of Splitrock. Losses were confined to young birds. By September 12, in an alfalfa field covering about 1 square mile, it was estimated that 400 young birds had died in a total population of 2,000. The cause of this loss was shown to be coccidiosis.

Scott (1940) reported another outbreak of coccidiosis occurring among sage grouse in Wyoming in 1939. This loss, which began in June and continued until September, again involved young birds. It was his opinion that the drought years concentrated the grouse on available water, thereby creating a situation favorable to the transmission of the coccidium.

Simon (1939) collected infected sage grouse near Battle Mountain and along the Sweetwater River in Wyoming. From his studies he concluded that there were two species of coccidia infecting the sage grouse: *Eimeria angusta,* described by Allen in 1934, and another which he described as *Eimeria centrocerci.*

TABLE 6

The Parasites of Sage Grouse (After Simon, 1940)

Endoparasites
    Protozoa (One-celled animals)
        1. *Eimeria angusta* Allen
        2. *Eimeria centrocerci* Simon
        3. *Tritrichomonas* sp.
    Cestoda (Tape worms)
        4. *Raillietina centrocerci* Simon
        5. *Rhabdometra nullicollis* Ransom
    Nematoda (Round worms)
        6. *Habronema urophasiana* Wehr
        7. *Cheilospirura centrocerci* Simon
        8. *Heterakis gallinae* (Gemelin) Freeborn
Ectoparasites
    Mallophaga (Biting lice)
        9. *Gonoides centrocerci* Simon
      10. *Lagopecus perplexus* Kellogg
    Acarina (Ticks and mites)
      11. *Haemaphysalis cinnabarina* Koch
      12. *Haemaphysalis leporis-palustris* Packard

Both of these species were reported to infect the caeca. A year later, Simon (1940) reported these parasites in sage grouse from Cody, Wyoming, and attempted experimentally to transmit the coccidia to pigeons and domestic chickens with no success. Honess (1942) reported the results of a group of experiments testing the possibility of cross infection between sage grouse and domestic chickens. All his attempts to infect the domestic chicken with viable cocysts from the sage grouse were unsuccessful.

During 1940 and 1941, field crews of the sage grouse survey examined 293 fecal droppings from various parts of the state. Eight per cent of these droppings contained coccidial cocysts. The positive droppings were from widely scattered areas.

From the information presented by Honess, it has been evident that infection with coccidia is common in sage grouse throughout most of the state. However, coccidiosis undoubtedly does not operate continuously as an epidemic disease. Honess stated that the disease ordinarily is associated with filth and wondered why it should be perpetuated in sage grouse which he considered to be a wide-ranging and seldom numerous species.

There has been speculation by Honess that coccidiosis in sage grouse may be associated with selenium poisoning, although the pathological condition observed in sage grouse with coccidiosis departed from that of chronic selenium poisoning in several important details. Based upon limited observations and only four cases of acute coccidiosis in isolated areas, he concluded that coccidiosis was still important as a factor in decimating sage grouse populations in localized areas.

Honess (1949) proposed a study of coccidiosis among sage grouse in Shirley Basin, Wyoming, during the summer of 1948. As it developed, no diseased birds were seen, although, based on analysis of droppings, nearly all of the adult birds were infected with coccidia. At this time additional studies were conducted by Honess on the occurrence of selenium and its possible relationship to sage grouse coccidiosis.

Knight and Beath (1937) named the Steele and Niobrara geological formations in Wyoming as bearing selenium. Both of these formations outcropped in the Shirley Basin area. The vetches (*Astragalus*) were prominently distributed over the entire area and were considered by Beath et al. (1939) to be some of the chief selenium indicator plants.

Studies of food habits material from two birds killed by cars in the Shirley Basin revealed that they were feeding extensively on sagebrush, vetch (*Astragalus grayi*), and red-legged grasshoppers (*Melanoplus femur-rubrum*). Tissues from two of the sage grouse, whose crop and stomach contents included the materials listed above, contained selenium in the following amounts:

	Adult Male	Young Bird
Liver and kidneys	10.6 ppm.	7.2 ppm.
Pectoral muscle	4.4 ppm.	7.2 ppm.

Similar studies were made on samples of sage grouse food, which were collected in the vicinity for comparison with the tissues:

Type of food	Ppm. selenium*
Astragalus grayi	819.0
Artemisia tridentata	2.0
Melanoplus femur-rubrum	39.2

*Average values

Honess assumes that sage grouse will tolerate greater amounts of selenium in their system than domestic animals (mammals), in which 5 ppm. will cause death. There has been some work done which suggests that the effects of selenium in the domestic chicken is first manifested by a decrease in the hatchability of eggs. No studies of this nature have been performed on the sage grouse.

Sage grouse were generally uncommon in the Shirley Basin during the summer season, although winter censuses and the presence of winter droppings indicated that the area was used by large flocks of birds during that season. This serves as an example of the importance of the various units of range in fulfilling the habitat requirements of the bird throughout the year.

As ordinarily constituted, diseased sage grouse were not commonly encountered during field observations in Eden Valley and adjacent areas from 1948 to 1951, although in mid-August of 1949 there was evidence that an epizootic of undetermined nature had affected a segment of the juvenile population. Diseased birds were observed, particularly in the vicinity of irrigation structures and alfalfa fields. These birds were weak, unable to fly any distance, generally exhibited the symptoms of partial paralysis, and had a diarrhetic discharge from the vent. In previous years several ranchers in Eden Valley had reported similar outbreaks, all of which occurred in the vicinity of the irrigation district. Freshly killed specimens of diseased birds were sent to the disease research laboratory for examination, and there was evidence of acute coccidiosis in several birds. The fact that this outbreak was associated with heavy concentrations of adult and juvenile birds in the vicinity of irrigation ditches seemed to suggest coccidiosis as the primary disease agent. Irrigation water probably acted as a vector of the disease, since droppings as well as dead birds contaminated the water as they were transported down the ditches. Losses primarily affected juvenile birds, although a few adult birds were afflicted. Outbreaks of this nature have been confined to the midsummer months and ordinarily were associated with irrigation systems or other artificially designed water structures.

Laboratory examination of an adult female found dead on a

nesting plot in Eden Valley revealed death to have been caused by the respiratory disease known as aspergillosis. This was apparently an isolated case and was not considered to be significant in management.

Bird lice of the *Mallophaga* groups were found on nearly every sage grouse examined on the study areas. Tapeworms, particularly of the genus *Raillietina,* heavily parasitized some birds. Many cases of these tapeworms protruding from the vents of sage grouse were noted while handling birds during trapping operations. In no case was any impairment of bodily processes noted.

Insect Control Programs

The widespread grasshopper control program launched in 1948 by the U. S. Bureau of Entomology and the Wyoming Department of Agriculture pointed out the necessity for conducting experimental studies on the effects of some new insecticides upon game birds. These studies were sponsored by the Wyoming Game and Fish Commission. The laboratory and field studies were assigned to George Post, assistant biologist, who has been responsible for the development of the project and who has presented the results herein summarized (Post, 1951b).

Determinations were made of the minimum lethal doses of Toxaphene and Chlordane on sage grouse, pheasants, and chukar partridge. These insecticides were to be used in grasshopper control work. In the case of sage grouse, 90 mg. (milligrams) of Toxaphene per kg. (kilogram) of body weight and 100 to 700 mg. of Chlordane per kg. of body weight appeared to be the minimum lethal doses. The results obtained on sage grouse were not conclusive, since it has been difficult to hold birds in captivity without early losses. Similar experiments on game farm pheasants revealed a MLD (minimum lethal dose) of Toxaphene to be 200 mg/kg of body weight and a MLD of Chlordane to be 500 mg/kg of body weight.

Post determined the histopathology of the game birds experimentally poisoned with Chlordane and Toxaphene, and then conducted field and laboratory studies on the effects of these insecticides on wild game birds encountered on the sprayed areas.

The areas selected for study were baited with Toxaphene and Chlordane administered as poisoned bran from planes at rates varying from 5 to 30 lb. per acre. Chlordane was mixed in a ratio of 0.5 lb. to 100 lb. of bran and Toxaphene at 1 lb. to 100 lb. of bran.

A total of 4,205,708 acres of Wyoming range land was treated with Toxaphene and Chlordane bran bait for grasshopper control during 1949 and 1950. The effects of the baiting operation on wildlife were observed by Post on 1,200 acres of baited land. Two hundred acres of unbaited land were used for control purposes. Eighteen dead and diseased birds which were found on baited plots proved to be affected by insecticide poisoning. One hundred and twenty-two other birds were found dead or affected on baited plots, but the cause of death or illness could not be diagnosed.

The total game bird mortality on baited plots was 23.4 per cent, whereas on the control plots, mortality was only 10.1 per cent. Sage grouse were the most numerous game birds on the baited plots and sustained a mortality of 21 per cent as compared with a negligible mortality on the control plots. Much of the dead material was unsuitable for laboratory examination due to the decomposition and disturbance by predators. In addition to the material listed above, many cases of dead birds were reported by ranchers in the area. The majority of local people contacted in the field reported finding some dead birds.

In an attempt to determine the extent of utilization of grasshoppers by game birds in the areas where insecticides were used, Post analyzed a series of game bird crops and stomachs. During the summer months, both adult and juvenile sage grouse consumed enormous quanities of grasshoppers; this food item occurred in 70 per cent of all material examined, and in some cases composed as high as 95 per cent of the total contents of a single crop.

An effort was made to determine quantitatively the amounts of Toxaphene present in dead grasshoppers collected from a baited area. An average of 28.69 mg. of Toxaphene was found in 100 gm. (grams) of freshly killed grasshoppers. This quantity of Toxaphene compared to 9.82 mg. of Toxaphene found in 100 gm. of grasshoppers which had weathered for two weeks.

Freshly killed grasshoppers in a baited area contained 28.69 mg. of Toxaphene per 100 gm. of live weight. The MLD of this insecticide for sage grouse appeared to be 90 mg/kg of body weight. The average summer body weight of over 200 adult male and female sage grouse in Eden Valley was 1,630 gm. Therefore, an average adult sage grouse would require about 147 mg. for a minimum lethal dose. In order to obtain this amount from grasshoppers alone, an adult sage grouse would have to eat approximately 506 gm. or slightly more than 1 lb. of freshly killed grasshoppers during a three-week period. Young birds would require smaller amounts for a MLD.

When Toxaphene was used in bran bait, as it was in Wyoming in 1949-50, it was possible to find dying and freshly killed grasshoppers lying about for extended periods after the bait was laid down. Birds could obtain recently poisoned grasshoppers for periods up to three weeks after the bait was spread. The cumulative effects of Toxaphene poisoning would make it possible for birds to pick up enough Toxaphene from grasshoppers to cause their death eventually.

Considering the tremendous land area involved in the grasshopper control program during 1949 and 1950, it was evident that large numbers of game and insectivorous birds fell victims to the lethal and sublethal effects of the insecticides Chlordane and Toxaphene.

The use of Toxaphene and Chlordane did not give satisfactory control of grasshoppers in 1949 and 1950. A new insecticide, Aldrin, was used for grasshopper control in 1951. Control measures consisted of spraying 2.0 oz. of Aldrin in kerosene per acre. Studies by Post (1951a) indicated that there was no appreciable difference in mortality of upland game birds on the sprayed areas as compared with the control plots, although excellent control on grasshoppers was obtained. Grasshoppers killed by Aldrin contained only 3.4 mgm. (micrograms) of the insecticide per 100 gm. of grasshoppers (live weight basis), as compared with 28.69 mg. of Toxaphene per 100 gm. in grasshoppers killed by the latter insecticide. Small amounts of Aldrin administered orally to pheasants in the laboratory caused a cessation of egg production.

Mechanical Injuries

Structural and functional changes produced as a result of physical injury to birds are classified as trauma. Included in causes of injuries of this type are predation, shooting, highway mortality, fighting, mowing machine casualties, freezing, flying into fences and wires, etc. Sage grouse frequently fail to recover from minor injuries, due to complications which normally follow.

Highway Mortality—Sage grouse mortality due to vehicular traffic occurred principally on paved highways which traversed areas heavily utilized for mating, nesting, and drinking purposes. Losses and injuries of this type were uncommon during fall and winter months, but were common in midsummer as the broods began to wander in search of food and water. In semidesert regions during morning and evening twilight, hens and young birds fed extensively on the luxuriant vegetation along highway right of ways. Mortality was heaviest during these periods and involved primarily hens and immature birds, although cocks were frequent casualties where strutting grounds were situated adjacent to traffic thoroughfares. Highway mortality afforded an index to sage grouse abundance, as evidenced by the results of studies to determine the extent of traffic pressure and sage grouse mortality along a stretch of main road which traversed the Dry Sandy—Pacific Creek study tract.

For purposes of determining sage grouse mortality due to highway traffic, periodic checks of road kills were made along a 15-mile stretch of the Lander-Farson state highway. A recording meter was utilized for obtaining traffic pressure over this route. During the period from April 1 to June 11, 1949, four sage grouse and fifty-nine jack rabbits were found dead along the right of way. A total of 12,138 cars traversed the 15-mile road strip during this period for an average of 172 cars daily. From August 14 to 26, fifty-nine jack rabbits and eight sage grouse were found killed along this same strip. A total of 3,395 cars traveled the route checked for a mean figure of 283 cars per day.

On September 22, a single check of highway mortality revealed the remains of ninety-six jack rabbits and six sage grouse within the 15-mile strip. Additional carcasses had been removed by predators or else obliterated by cars, since the previous inspection on August 26.

Sage grouse highway mortality was highest during the months of June, July, and August. Jack rabbit and cottontail mortality occurred during periods of darkness. Antelope sustained fairly heavy losses along trunk highways in Wyoming during the late summer months. All of these animals were attracted to the areas bordering highways by an abundance of roadside vegetation. A variety of valuable forage and cover plants such as rabbitbrush, Russian thistle, sweet clover, alfalfa, and some grasses thrive along highways during the dry summer months due to the increased moisture, soil fertility and cultivation associated with road maintenance. Most of Wyoming's highways are of oiled or blacktop construction which renders difficult the motorists' problem of noticing sage grouse during twilight periods. Long and relatively straight stretches of highway are common in the sagebrush prairie regions. This results in extremely high automobile speeds, further endangering sage grouse and other important game species. Several instances were observed wherein sage grouse crashed through the windshields of speeding cars, resulting in injury to the occupants.

In addition to the plants which naturally become established along Wyoming highways, the State Highway Department has planted several forage plants, e.g., sweet clover, bromegrass and crested wheatgrass, as an aid in preventing erosion and stabilizing the shoulders. These plants lure sage grouse, antelope, rabbits and deer onto the highway proper. The elimination of such food and cover plants from the immediate vicinity of federal and state highways would serve to reduce wildlife mortality on Wyoming highways. It would be desirable from the wildlife standpoint to create by chemical means, a permanently sterile strip of ground adjacent to the road pavement. Such a procedure would be consistent with the present highway department policy of maintaining clean and deep borrow pits as an aid in preventing snow accumulation on trunk roads within the state. In order to reduce the drifting of snow, considerable time and money are expended prior to the snow season in destroying and removing the annual accumulation of vegetation bordering the highways.

The Wyoming Highway Department has expressed a sincere interest in this problem of highway mortality to wildlife and has shown a willingness to co-operate with the Game Department in an attempt to develop a program of roadside improvement which will reduce the annual slaughter of wildlife on the state's road system, and at the same time fulfill its own objectives in construction and maintenance. It is recommended that such a study be inaugurated.

Mowing Machine Casualties—During July, August, and September sage grouse drifted into sagebrush areas adjacent to Eden Valley ranches for the purpose of feeding on alfalfa crops. The usual feeding periods for adult birds was in early morning and late afternoon. Retirement to sagebrush areas was customary for adult birds during midday and at night. However, females with young tended to remain in alfalfa fields throughout the day and suffered heavy losses from mowing operations during the harvest season, due to their habit of squatting and hiding at the approach of farm machinery. At the altitudes prevailing in Eden Valley, two crops of alfalfa normally are harvested each growing season. Flocks of from 200 to 300 hens and juveniles were observed in many alfalfa fields during haying operations. Estimates made by mowing machine operators indicated that juvenile birds suffered substantial losses. Nests were only rarely placed in cultivated fields, and in any event, nesting activities would be completed before the onset of the alfalfa harvest season.

Miscellaneous Injuries—There is evidence that birds occasionally sustained frozen toes during the winter months. A very small percentage of the grouse handled during the course of trapping operations exhibited loss of an entire toe or a segment thereof, but were otherwise perfectly healthy and normal. Two cases were noted of adult cocks with only one foot. A bird with one foot was observed to be strutting. Shooting casualties have caused losses of limbs and toes. Crippling losses during legal hunting seasons were presumably low, due to the nature of the cover and the inability of wounded sage grouse to escape, as compared with a game bird such as a pheasant. Territorial conflicts between adult cocks on the strutting grounds frequently were violent enough to produce minor injuries to the combatant's wings. An adult cock which apparently had suffered a broken wing as a result of such actions was collected on a strutting ground.

Rodent Control Programs

There has been a widespread belief that rodent control programs over western range lands have been an important factor in the decline of sage grouse numbers in some areas. It has not been possible to gather any first-hand data on this supposition, due to the absence of rodent control programs on the areas during the period of study. Girard (op. cit.) was unable to locate a single dead sage grouse on a 3,500-acre tract thoroughly poisoned for ground squirrels in mid-July in Sublette County, Wyoming. Oats were used for bait and it was assumed that strychnine was used as the poisoning agent. On a section of one ranch were ground squirrels were abundant and sage grouse plentiful, the latter withdrew from the control areas as the exposed carcasses of the dead squirrels began to decay and did not return for several days.

The Influence of Sage Grouse on Its Environment

Physical Environment

In common with larger game species and livestock, sage grouse are capable of exerting a noticeable effect upon their environment. The normally slight influence expected as a result of their small size and extreme mobility is frequently offset by the increased pressures arising from their abundant numbers and habits of concentrating in relatively small areas. This is especially the case in the vicinity of irrigated areas, water holes, and strutting grounds. On these areas, intensive daily use for several months results in marked effects upon their habitat, as indicated by the heavy utilization of plant parts for food, trampling of vegetation, elimination of plant cover, incorporation into the soil of material from droppings, and by consumption of insects.

Sage grouse droppings may be found scattered throughout the sagebrush prairie. They are ordinarily interspersed with the fecal remains of jack rabbits, domestic sheep, and antelope. Due to the nature of the climate these droppings remain intact, and can be identified for several years following their deposition. Nitrogenous compounds, crude fiber, and minerals are continually being returned to the soil from animal wastes.

During the course of some three months of continuous strutting activity each year, the soil on individual strutting grounds becomes well-loosened and stirred, facilitating invasion by grasses and forbs during the summer months.

Plant Cover

Many strutting grounds exhibit a central area which is permanently barren of shrubby plants due to continuous use of the area by sage grouse during the courtship period. On newly established strutting grounds the shrubs gradually become stunted through the years and eventually will be displaced by native grasses and forbs. Radiating from the barren area, or in its absence from the center of all strutting grounds, is a well-defined zone of overbrowsed sagebrush which extends to the outer limits of the area utilized for daily strutting and assembly. Prior to their departure from the strutting grounds, the birds engage in feeding activities. Similar zones of vegetation indicative of overuse by sage grouse are found in areas surrounding water holes and in alfalfa fields supporting large concentrations of birds. During most seasons of the year sage grouse are sufficiently dispersed to cause little or no damage to vegetation from their feeding and trampling activities. In alfalfa fields thrust into the center of extensive sagebrush areas the opposite situation exists. Hordes of sage grouse frequent these fields twice daily for purposes of feeding on the alfalfa and drinking the water from irrigation systems. When these flocks number several hundred birds, as was the case in Eden Valley, reduction of the crop from feeding and trampling reaches limits intolerable to the rancher. I have determined from crop contents studies that one thousand adult birds feeding twice daily in alfalfa fields would consume or otherwise destroy in a six-weeks' period approximately five tons of alfalfa hay (air-dry basis). Heavy damage to alfalfa hay has occurred in Eden Valley for the past three years with large concentrations of sage grouse moving into alfalfa fields during the months of June, July, August, and September. Ranchers in other agricultural areas of the state have reported sage grouse damage to bean, pea, alfalfa, and clover crops, but these depredations were of less intensity than those recorded in the Eden Valley hay fields.

Animals

In view of their propensity for consuming large numbers of insects during the spring and summer months, it would seem that the presence of sage grouse in agricultural districts has not been entirely detrimental to the ranchers. Post (1950a) has emphasized the value of sage grouse in eating large quantities of grasshoppers in plague areas of the Powder River Basin in Wyoming. Farmers on the Riverton reclamation district were greatly relieved to observe the large amounts of insects contained in crops of sage grouse collected in their bean fields by a deputy game warden. Insect plagues have only infrequently been recorded in agricultural and range areas of the Green River Basin, where the sage grouse occurs abundantly.

There was every indication that the golden eagles, associated with winter concentrations of sage grouse, followed the grouse migrations down drainage in the autumn, and moved into the foothills simultaneously with the chickens in the spring.

During severe winters with a heavy snowfall, small rodents were largely unavailable to coyotes as a source of food. Consequently, there was a movement of coyotes, as well as eagles, from the mountains onto the semidesert where sage chickens, jack rabbits, antelope, and mule deer were concentrated on their winter ranges. Jack rabbits and cottontail rabbits were considered to be the primary prey species of golden eagles and coyotes on the sagebrush winter range. Sage grouse were a secondary, though important, prey species for the eagles. No other important avian or mammalian predators were associated with the grouse during the winter period. Badgers, ground squirrels, and magpies were present on sage grouse nesting areas. As predators upon sage hens, they functioned chiefly as nest destroyers. They were not directly dependent upon the birds themselves as were the golden eagles. Neither were the nest predators transitory, in the sense that they followed grouse flocks during their seasonal migrations.

PART II

—

THE NATURAL HISTORY OF SAGE GROUSE POPULATIONS

INVENTORY AND CENSUS METHODS

Reconnaissance Methods Based Upon Recognition of Sign

The recognition of animal sign has always been an essential part of the woodsman's craft. The ability to properly interpret sign, be it Indian or game, was an attribute highly valued in the scouts who led early immigrant and exploratory expeditions. The hunting guide possesses keen powers to detect, observe, and interpret animal sign in the field. Similarly, a great deal of information on the distribution and relative abundance of sage grouse can be realized from the mere ability to identify and evaluate sign left by the birds in the fields. It is in reality a simple method of reconnaissance which can be utilized for inventory purposes in the absence of more organized census techniques.

Tracks

Sage grouse make a chicken-like track when walking. The hind toe is diminutive, slightly elevated, and does not leave an imprint in the footfall. The impression left by the middle toe is noticeably longer than that from the outer or inner toe. Due to differences in size of the male and female foot, especially in length of the middle toe, the track of a mature sage grouse may be used as an indication of sex. The length of the middle toe on adult males will average about 66 millimeters, and on twenty-five birds examined, the middle-toe length ranged from 62 to 71 mm. The middle toe of fifty-seven adult females ranged from 48 to 61 mm. and averaged 55 mm. In applying this relationship to actual tracks in the field, the distance from the tip of the middle-toe print to the heel of the footprint should be measured. Sage grouse take short strides, and at a normal gait the distance between footfalls seldom exceeds the length of an individual footprint. If the bird is walking in loose sand or snow, the middle toe is dragged, leaving a mark between successive footsteps. The frequency of snowfalls in the winter and the prevalence of sandy soil types over large parts of the birds' range in summer greatly facilitate track reconnaissance. A maze of graded and unimproved roads, sheep wagon trails, and dry stream beds furnishes sandy mediums for the implantation of sage grouse footprints. If sage grouse are present on an area, their tracks will be evident at these locations.

Droppings

Along with tracks, sage grouse droppings provide an index to the distribution and relative abundance of the species. These droppings survive the weathering processes for indefinite periods of time and furnish good evidence of grouse occupancy.

Strutting Grounds

During the period of the year when strutting grounds are temporarily abandoned, they may be distinguished by the abundance of sign deposited by the grouse during their courtship activities. The most striking feature noticed on the majority of display areas is the central clearing surrounded by a zone of stunted and partially denuded vegetation. Within the display area may be found a three-months' accumulation of intestinal droppings, interspersed with the solidified remains of cecal droppings. Loose feathers may be found, depending upon the length of time since strutting activities.

Watering Places

Throughout the semiarid prairie and foothills region, springs, seeps, and water holes supplement streams as sources of drinking water for sage grouse. The flocks tend to concentrate around these locations in late summer. Sage grouse tracks, droppings, and feathers give evidence of utilization of these watering areas.

Sage Grouse Census Techniques

Strutting Ground Census

Locating Strutting Grounds—For purposes of courtship displays and mating performances, sage grouse assemble during the breeding season in groups of varying size on permanently established strutting grounds. The three species of prairie grouse (pinnated grouse, sharp-tailed grouse, and sage grouse) are polygamous and exhibit similar breeding behavior on "ancestral" display grounds in successive breeding seasons. Barring total annihilation of all males utilizing a display area or complete obliteration of the physical aspects of the strutting area itself, generation after generation of birds characteristically utilize the same parcel of land for the annual breeding ritual. The duration of occupancy of individual grounds certainly extends over many years, although in any limited period new grounds are created by young birds, and others probably are passing out of existence due to the disintegration of a localized male population.

Several biologists have established the reliability of census methods associated with auditing the sounds produced by upland game birds during courtship performances. They have developed efficient and practicable census techniques that are likewise useful in the determination of grouse, quail, and pheasant populations. Hamerstrom (1940), Lehmann (1941), Schwartz (1945), and others have utilized the same principle in their prairie grouse studies. Kimball (1949) explored the field of audio-censusing of game birds and summarized the development of the pheasant crowing-count census now being used extensively to determine the distribution and yearly trends exhibited by South Dakota pheasant populations. The audition of sounds produced by sage grouse in their courtship displays has also been a feature of censuses designed to locate sage grouse strutting grounds. The nature of the terrain and vegetation over sage grouse range has made it possible for humans to utilize their own sense of sight as a supplemental factor in locating the strutting males and hence the strutting grounds.

The lengthy strutting season, lasting as it does for approximately three months, provides an opportunity for a small field crew to census several hundred square miles for strutting grounds in a single season. On sage grouse range in Wyoming one experienced man operating independently can effectively cover 200 square miles in a single strutting season.

Censusing of Cocks—The censusing of a large area of land to locate all strutting grounds is a distinct operation and should not be confused with the type of census which is concerned with enumerating male birds on individual strutting grounds. This latter census must be confined to certain dates during the seasonal strutting period and at particular hours on the days specified. Censuses of cocks on strutting grounds will accurately represent the total numbers of males in any given district, provided that the counts are conducted on a unit-area basis involving 100 per cent coverage of a large land sector. Censuses of males on individual grounds selected at random over a segment of sage grouse range will afford the most practical method of obtaining indices to yearly population trends.

The task of locating strutting grounds and censusing total male populations on an extensive area of land in the Green River Basin is facilitated by the following factors:

1. Extensive flat to rolling prairie characterized by small buttes, ridges, and knolls which are useful as points of vantage.

2. Low, flat, and homogeneous vegetative cover types of sagebrush and associated plants, facilitating distant observation of cocks in full display.

3. Clear atmosphere, low humidity, and lack of wind during morning and evening twilight hours.

4. Males' habit of utilizing the same strutting grounds in relatively constant numbers throughout a lengthy breeding period.

5. Tendency of males to select clearings and open areas in the brush for strutting activities.

6. Large size of the cocks and their spectacular courtship display in a brilliant white plumage.

7. Intensive and active courtship displays little affected by close range observation from a vehicle.

8. The distinctive booming sound produced by strutting males which is audible at distances of from 1 to 3 miles under ideal weather conditions.

9. An excellent network of federal, state and county roads, supplemented by a maze of sheep wagon trails, making possible a rapid and thorough coverage by truck.

In reporting the results of the Colorado sage grouse survey, Keller et al. (1941) concluded that strutting ground censuses were unreliable as indicators of population trends because of daily fluctuations in numbers, although they considered them to be fairly reliable for data on sex ratios and breeding activity.

Sage grouse strutting ground censuses in Wyoming proved to be very accurate in representing male populations and were invaluable as indices to yearly population trends. The same strutting grounds were utilized year after year and in a definite weekly pattern during the breeding season. Disturbances by natural enemies and humans, unfavorable weather, and the irregular appearance of juvenile males all caused fluctuations in the daily patterns of occurrence. Censuses should be conducted at daily intervals during the three-week period immediately following the peak of mating activity. On areas studied in the Green River Basin at elevations from 6500 to 7500 feet, this peak occurs during the third week of April. Actual observations should be made from a motor vehicle at extremely close range on clear, still mornings, and confined to the one-half hour period prior to sunrise and the one-half hour period immediately following sunrise. The final census figure for individual strutting grounds should be derived from the maximum count obtained from censuses of equally high intensity taken on a minimum of three different days, not necessarily in succession. This will offset the daily variations in numbers of cocks which frequently are encountered in the field. In order to obtain reliable information on total populations, the censuses should be conducted on a fairly large land area basis with a 100 per cent land coverage, at least during the initial year of investigation. For purposes of management, the yearly population trends are considered reliable when based upon a representative sampling of grounds located in previous years.

Strutting ground censuses are not reliable for obtaining total female populations due to the fluctuating pattern of visitation and the difficulty of observing females. Sex ratios based upon strutting ground censuses are considered inaccurate for the same reasons.

Nesting Census

The nesting census was accomplished by spacing four men 15 feet apart and systematically traversing individual plots, utilizing guide flags for alignment purposes. Even spacing between members of the census crew was maintained by use of a length of rope from which strips of wooden lath were suspended at 1-foot intervals. On several occasions, the lath strips actually touched incubating females without flushing them. In no observed case was a lath directly responsible for flushing a hen, although it was felt that the noise associated with its dragging over the brush assisted in frightening hens from their nests. Each man in the crew was responsible for visual detection of nests and incubating females for a distance of 7½ feet on either side of his station. Hens normally would flush from nests upon the approach of the crew, although during the late stages of incubation they invariably remained upon nests while the census crew passed. The nature of the nesting cover afforded opportunities for easy detection of incubating hens by experienced crewmen. Four men could effectively cover 10 acres of nesting habitat per hour, or approximately 80 acres per day. Individual nests were marked by placing a lath 50 feet from the nest site.

Plots of 40 acres were selected as sampling units. Within the

Eden Valley and Dry Sandy—Pacific Creek study areas, these plots were located in a standardized random pattern at approximately township intervals (Figures 3 and 4). Twenty of these 40-acre plots were on the Eden Valley area and fifteen were situated on the Dry Sandy—Pacific Creek tract. For convenience in censusing, the 40-acre plots were divided into 10-acre subunits.

Sage grouse nesting censuses should be delayed at least one week after the laying of the first full clutches of eggs, in order to minimize the possibility of nest desertion upon the part of the hen. In Wyoming, nesting activity commences on or about April 20 and generally terminates by July 1.

Sage Grouse Brood and Flock Census

The observation and censusing of sage grouse broods are relatively simple field procedures as compared with brood studies of most other upland game birds and waterfowl. This is due to the nature of the protective cover and the behavior of the birds. The open character of sagebrush types on all but the higher foothill ranges affords little real escape cover for sage grouse broods once they have been noticed. The majority of hens exhibit great tolerance to observation of their broods from a car. The relatively few hens which react to disturbance by vehicles generally respond by causing their broods to remain motionless on the ground rather than to scurry off to cover. Patience on the part of the observer will be quickly rewarded by a resumption of normal activity on the part of the brood. Hens and their broods rarely flush unless the observer presses them on foot or unless the truck happens to move suddenly into their midst. The majority of brood counts will be complete, in the sense that all the chicks can be observed. Occasionally the brood censuses will provide only partial counts, due to heavy cover, distance of observation, or alertness of the hen in scattering her young.

Censuses of adults and broods in this study were initiated during the summer months of 1948. They were resumed on a minor scale during the summer of 1949, and were discontinued completely during 1950, as a result of the writer's participation in trapping and transplanting operations the two latter years. Censuses were commenced in mid-June and were continued throughout July and August. Brood censuses should not be commenced until young birds are at least three weeks old; prior to this time, the chicks are under close supervision of the hens, and are kept secluded. Brood censuses should not be conducted after mid-August, since field observations indicated that most family organizations commence to disintegrate once the young attain ages of ten to twelve weeks.

Brood and adult counts were conducted by observation of roadside areas from a half-ton pickup truck at intervals approximately one week apart, from standardized routes in the Dry Sandy—Pacific Creek region. These roadside censuses provided information on sage grouse nesting success, brood survival, brood movements, and the location of optimum nesting areas. Censuses were made during the late afternoon and early evening in order to coincide with the time of greatest sage grouse activity. Roadways utilized in the census were usually located in close proximity to streams, springs, and moist wild hay meadows, which were focal points for brood and adult concentrations during evening periods. As pointed out earlier, graded roads provided excellent dusting areas for young birds, and paved roads furnished quantities of green food along their right of ways. All these factors tended to create abnormally high numbers of broods and adult birds along the census routes. Consequently, it was difficult to ascertain the number of birds per unit area from the route-census figures alone. Darkness and road conditions occasionally prevented coverage of the entire route. The number of broods encountered on each census varied with the weather conditions and the time of passage through concentration areas.

Roadside censuses were normally limited to distances of 100 feet or less on each side of the road. In 1949 the censuses were taken by one observer, an arrangement which prevented a thorough coverage of areas on both sides of the car. Since two observers were used on all census trips in 1948, the figures for that year have been reduced by 50 per cent for comparative purposes.

In both years, flocks of from eighteen to thirty-five hens and young were commonly observed in late August and early September. From mid-September until November, it was difficult in the field to distinguish juvenile birds of either sex from the adult hens.

When conducted in accordance with standardized methods and along roadsides from a motor vehicle, brood and adult censuses of sage grouse are considered to give fairly accurate indices to nesting success, juvenile mortality, and annual productivity. For management purposes they are superior to nesting censuses, since much larger units of the birds' range can be sampled and more birds observed with substantially smaller field crews.

Use of the Aerial Survey in Censusing Sage Grouse

Sage grouse wintering areas are frequently snowbound, thus preventing effective survey by surface transportation. On these occasions reconnaissance of winter concentrations can be accomplished by the use of aircraft. In those states utilizing aircraft for winter censuses of antelope herds, the location of sage grouse wintering areas may be ascertained as a part of the operation, since both species ordinarily exist together on the winter ranges. Low-flying aircraft generally flush sage grouse flocks along the path of flight, and sage grouse tracks in the snow are easily detected when flying at low altitudes. Although accurate counts are not possible from the air, the aerial survey does have definite usefulness in determining the general region of the birds' winter range. It has been suggested that planes might be used in locating sage grouse strutting grounds although this technique was not attempted during this study. Aerial detection of sage grouse would seem to be difficult in the absence of snow cover.

LIFE HISTORY OF THE SAGE GROUSE

Nesting

Reproductive Success

Nesting studies were made on the Eden Valley irrigation district and the Dry Sandy—Pacific Creek area as previously defined. Fourteen hundred acres, representing approximately 1 per cent of the total acreage on both study areas, were censused in 1949 and again in 1950 with 100 per cent coverage each year. Nesting densities and nest mortalities for each of the two major study areas appear in Table 7.

The nesting census was begun May 7, approximately one week after the laying of the first complete clutches of eggs. It was finished on June 4, more than two weeks after the first nest hatched. The hatching period commenced during the third week in May, with the peak of hatching occurring the last week of May.

The ultimate fates of all nests studied in 1949 and 1950 are shown graphically in Figure 6. Eden Valley plots (800 acres) contained fifty-four nests of all categories in 1949, as compared

98

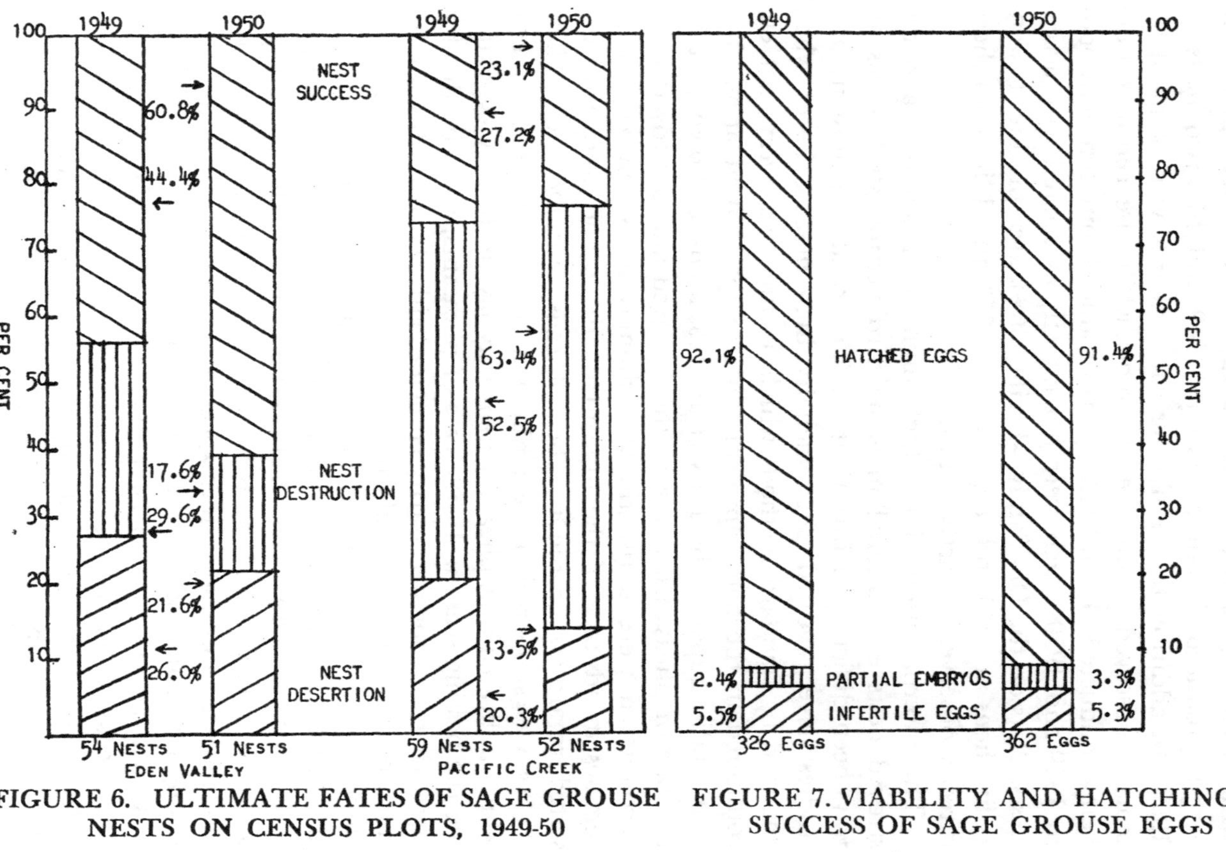

FIGURE 6. ULTIMATE FATES OF SAGE GROUSE
NESTS ON CENSUS PLOTS, 1949-50

FIGURE 7. VIABILITY AND HATCHING
SUCCESS OF SAGE GROUSE EGGS

with fifty-one nests in 1950. Dry Sandy—Pacific Creek plots (600 acres) produced fifty-nine nests in 1949 and fifty-two nests in 1950. In addition to the nesting data, records were kept of all females observed on the nesting plots while the censuses were being conducted. Eden Valley plots produced fourteen hens both in 1949 and 1950 which were not directly associated with active nests. Plots on the Dry Sandy—Pacific Creek area produced twenty-five hens in 1949 and forty-seven in 1950, in addition to those directly observed on active nests.

The interpretation of the high number of deserted and destroyed nests encountered in game bird nesting studies has always been difficult for investigators. The results of most nesting studies have revealed phenomenally high nesting losses when analyzed by standard methods which assume that every nest encountered in the field represents an attempt at incubation. Recent studies reported by Buss et al. (1951) have shown that in the case of pheasants, many of the deserted and destroyed nests encountered were unquestionably dummy nests constructed by females in advance of the urge to incubate, and were thus deserted regardless of the number of eggs initially laid. Their theory was supported by examination of the ovulated follicles which indicated that pheasant hens may lay up to fifty eggs during the nesting season. In the case of pheasants, the stimulus for egg-laying did not necessarily coincide with the stimulus for incubation. In any event, pheasant hens were capable of and actually did lay several clutches of eggs each year, either as a result of normal behavior or of stimuli caused by initial disturbance or destruction of their nests. Considering the length of their potential breeding periods and the nature of their courtship activities, repeated nesting performances seemed more likely to occur among pheasants than among sage grouse.

As will be outlined in later discussions, the courtship period and breeding behavior of the sage grouse precludes the possibility of repeated nesting attempts on the part of hens throughout the spring and summer months. The phenology of mating, egg-laying, incubation, and hatching for this bird followed a prescribed pattern of occurrence, which did not vary appreciably throughout the years of this study.

There was every indication that the incubation period commenced immediately after the complete clutch of eggs was laid. Although the actual nesting census was delayed until May 7 in order to prevent undue disturbance of hens during the egg-laying period, a random search for nests was initiated at a period corresponding with the peak of mating. In this manner, the complete histories of several dozen nests were obtained. It seemed unlikely that female sage grouse established dummy nests with the same degree of regularity displayed by pheasants.

The frequency of hatching by weekly intervals for all nests observed in 1950 is shown in Figure 8. No records were obtained of nests hatching later than the last week in June. Since strutting activities generally had ceased by late May, this would seem to provide considerable evidence that the duration of sage grouse mating and nesting behavior is regulated by the length of the strutting cycle in the male.

In computing nesting populations, renesting attempts and nests missed by the census crew introduced sources of error. During the 1949 census, the inner guide man on the line of advance reconnoitered the area already searched by the outer man on the previous trip. This system of double-checking turned up three nests which had either been deserted or were in the process of incubation, and which had been missed by the first man over the strip. Since the lanes covered by the two men in the center of the line were only checked once, it was assumed that a similar number of nests had been missed by each; therefore, in computing total numbers of nests, the figure 9 was added to the number of nests actually counted on the entire 1,400-acre study area. In 1949, a total of seventy-five undestroyed nests (in process of incubation or else deserted) were found at the time of censusing the nesting plots. Hence, it was estimated that about 12 per cent of the nests were missed during the census.

Late hatching dates, new nests adjacent to destroyed or deserted nests, and clutches with less than six eggs were considered to be representative of renesting attempts. On this theoretical

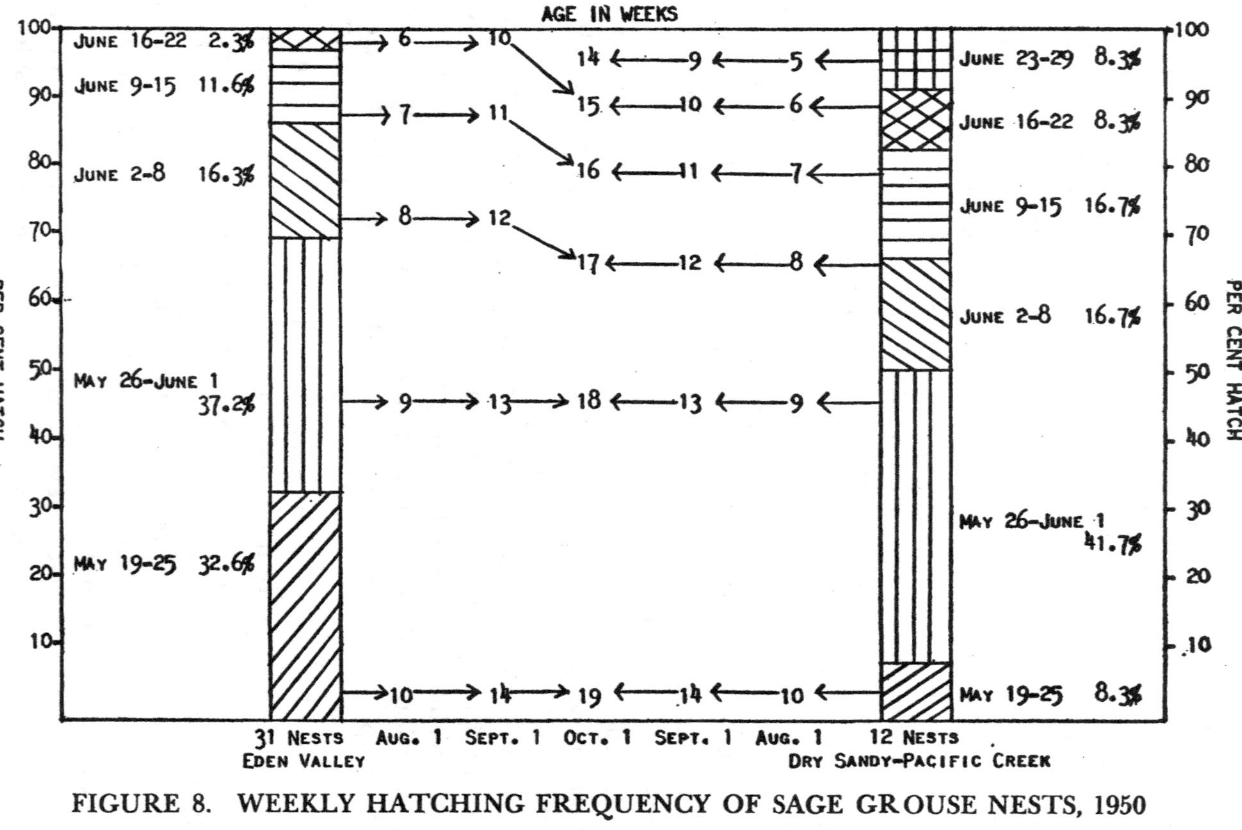

FIGURE 8.   WEEKLY HATCHING FREQUENCY OF SAGE GROUSE NESTS, 1950

basis, five nests were classified as renests; of this number, four hatched successfully between June 7 and June 20. Since 35 per cent of all nests located in 1949 hatched successfully, it was computed that twelve nests, or approximately 10 per cent of the total, represented renesting attempts.

For the purposes of computing female numbers, the errors introduced from nests overlooked by the censusing crew and from hens renesting were assumed to be compensatory. If this assumption is allowable, no corrections need to be applied to the original nest-density figures. All field observations indicated that the entire female population conducted at least one nesting attempt. For instance, during late April and early May no hens were found assembled in flocks which were so commonly observed after the middle of May. Neither were there any data to indicate that immature hens had failed to nest at the age of one year. Microscopic examination of the ovaries of three yearling hens collected on June 14 disclosed the presence of ovulated follicles, indicating that ovulation had recently occurred. Extensive brood censuses conducted by car during June, 1948, on the Pacific Creek area revealed that 43 per cent of all hens observed were with young, whereas 39 per cent of all nests located on the same area in 1948 had hatched successfully. During the 1949 and 1950 censuses, records were kept of all females observed on nests and also of all other hens not directly associated with nests. Females which flushed in flocks of two or more birds were assumed to be finished with nesting activities for the season and as such represented hens from destroyed or deserted nests. On the Dry Sandy—Pacific Creek plots, censused in late May and early June both in 1949 and 1950, ten flocks of hens totaling seventy-two birds, and varying from two to nineteen birds per flock, were flushed from nesting plots. These hens represented 66 per cent of all females observed on the plots during the course of the study. This percentage figure compared closely with the combined destruction and desertion of 74 per cent for the same area during those years.

The Eden Valley plots were censused from May 8 to May 25. On these plots the females were consistently flushed as singles.

On two occasions, flocks of three and five birds were flushed. On May 19, 21, and 22, flocks of six, five, and four females, respectively, were flushed in Eden Valley but not from established nesting plots. This was the first evidence of hens assembling in flocks of any size. At this time both destroyed and deserted nests were being found in increasing numbers.

TABLE 7
Nesting Success Based on Number of Nests on Census Plots—
1949-1950

	Acres	Total Nests	No. Nests and Per Cent Destroyed	No. Nests and Per Cent Deserted	No. Nests and Per Cent Hatched
Eden Valley					
1949	800	54	16—29.6%	14—26.0%	24—44.4%
1950	800	51	9—17.6	11—21.6	31—60.8
Area totals	1600	105	25—23.8	25—23.8	55—52.4
Pacific Creek					
1949	600	59	31—52.5	12—20.3	16—27.2
1950	600	52	33—63.4	7—13.5	12—23.1
Area totals	1200	111	64—57.7	19—17.1	28—25.2
Grand totals	2800	216	89—41.2%	44—20.4%	83—38.4%

In 1949, the number of nests in all categories exceeded the total number of females found on both Eden Valley and Pacific Creek nesting plots. In 1950 the reverse was true; the total number of females on plots greatly exceeded the number of nests. In order to give a simpler and perhaps more accurate analysis, the nesting data for 1949 and 1950 are combined in Table 8. Destroyed and deserted nests have been further classified into nests originally found in a destroyed or deserted condition and those which were destroyed and deserted subsequent to location. In this method of analysis, 105 nests and 101 females were located on Eden Valley nesting plots. On the Dry Sandy—Pacific Creek plots 111 nests and 109 females were found. By basing nesting success on the total number of females encountered on nesting plots rather than on the total number of nests, a slightly higher nesting success was obtained. The small incidence of renesting attempts would tend to further widen the disparity between total nests located and total hens encountered. A few of the stray hens flushed (10 per cent or so) undoubtedly built nests on the nesting plots subsequent to the actual census. One nest of this kind with five eggs was found on a plot following the initial census. In view of the possibility of limited occurrences of renesting and the laying of an occasional dummy nest, it was decided to use the total number of females encountered on the plots, rather than total nests found, as the index to nesting success and female nesting densities (Table 8). On this basis combined nesting success in Eden Valley in 1949 and 1950 was 54 per cent as compared with a 26 per cent nesting success on the Pacific Creek area.

Nest destruction accounted for only 23 per cent of Eden Valley nests as compared with 57 per cent of the Pacific Creek nests. Since the latter area was censused at a later date than the Eden Valley plots, it may be assumed that many of the nests listed as being destroyed were originally deserted, and subsequently destroyed prior to the census. For example, in 1949, all nests which had been found deserted in April and May were rechecked in late June. At this time 31 per cent of all nests formerly deserted had been destroyed by predators.

The higher incidence of nest destruction on the Dry Sandy—Pacific Creek plots was further substantiated by the number of old nests encountered on the two areas. In 1949, 173 old nests were found in Eden Valley as compared with 90 old nests found on Pacific Creek plots. The significance of this data lies in the

TABLE 8

Nesting Success Based on Number of Hens Flushed on Census Plots—
1949-1950

Eden Valley Plots

Hens flushed from active nests	73
Total hens flushed	101
Total nests found	105
Nests hatched and per cent hatched	55—54%
Nests destroyed when found	9
*Total nests destroyed and per cent destroyed	23—23%
Nests deserted when found	19
*Total nests deserted and per cent deserted	23—23%
	100%

Dry Sandy—Pacific Creek Plots

Hens flushed from active nests	37
Total hens flushed	109
Total nests found	111
Nests hatched and per cent hatched	28—26%
Nests destroyed when found	56
*Total nests destroyed and per cent destroyed	63—57%
Nests deserted when found	16
*Total nests deserted and per cent deserted	18—17%
	100%

*Figures reduced by proportional amount from listings presented in Table 7, in order to base percentages on total number of hens, rather than on total number of nests.

fact that in nests which hatch normally, the shells remain intact and visible for several years, whereas in the case of destroyed nests the shells are badly broken, scattered, and in many cases buried. Evidence of a sage grouse nest was seldom visible the year following its destruction.

Destruction and Desertion of Nests

Ground squirrels, badgers, and magpies were responsible for most of the nest destruction in 1949 (Table 9). The determination of the causes of nest destruction was complicated due to the likelihood of additional disturbance of nest and shells by other predators following the initial act of destruction. Nest disturbance after desertion presented an additional hazard in the correct evaluation of nesting mortality. The assignment of nest destruction to a specific predator was based on a knowledge of that animal's existence on the nesting area and a complete examination of animal sign observed in the vicinity of each destroyed nest. After a careful examination of many cases of nest destruction and a checking of examples of nest predation recorded in the literature, it was possible to identify the predator with a high degree of accuracy.

TABLE 9

Sources of Sage Grouse Nest Destruction—1949

Eden Valley Plots

Type of Predator	Nests Destroyed	Per Cent Destroyed by Each Predator
Badger	7	44%
Ground squirrel	5	31
Magpie	3	19
Coyote	1	6
	16	100%

Dry Sandy—Pacific Creek Plots

Ground squirrel	15	48%
Badger	10	32
Coyote	4	13
Magpie	2	7
	31	100%

Total Percentage of Nest Destruction on Both Areas

Ground squirrel	20	42%
Badger	17	36
Magpie	5	11
Coyote	5	11
	47	100%

Excluding the ratio of nest destruction by magpies, no significant differences in types of nest predation were noted on the two areas. Magpies are not commonly found on the Dry Sandy—Pacific Creek area due to the absence of ranches and willow thickets. It was felt that the factors of civilization associated with irrigation systems and ranching were responsible for smaller numbers of mammalian predators and an accompanying lower rate of predation upon sage chicken nests in Eden Valley.

Three species of ground squirrels occurred on the nesting plots: the Wyoming ground squirrel (*Citellus richardsoni elegans*), the small striped ground squirrel (*Citellus tridecemlineatus parvus*), and the least chipmunk (*Eutamias minimus* ssp.). More than 40 per cent of the nest destruction was caused by ground squirrels, presumably the first two species listed. Nests destroyed by ground squirrels may be completely robbed, leaving no trace of eggs, or the eggs may be eaten at the vicinity of the nest. Eggs usually were opened at the end, although the sides were occasionally broken. Holes were typically as large as the egg diameter with the edges curled inward. Tooth marks occasionally were present; the shells were seldom crushed; and the nest lining was normally undisturbed. Deputy Game Warden Eugene Knight reported observing ground squirrels (probably *Citellus armatus*) destroying sage grouse nests and eating eggs in the vicinity of Cokeville, Wyoming in 1949, and reportedly found several nests exhibiting a similar type of destruction. Other investigators have recorded instances of ground squirrels destroying the nests of ground-nesting birds (Sowls, 1948). While conducting sage grouse nesting studies in Colorado, Keller et al. (op. cit.) kept a record of the animals which trap, poison, and hair sample evidence showed had visited a series of dummy nests set out in sage grouse nesting areas. A total of ninety-seven visitations was recorded; of this number, 76 per cent were made by ground squirrels and 15.5 per cent by badgers. Most ground squirrel records were obtained in traps. These workers in Colorado concluded that ground squirrels are not a factor in actual nest predation but bother eggs more from curiosity than from any intent to destroy. In opposition to this conclusion was their statement acknowledging that ground squirrels are guilty of moving eggs and of digging in nests which they indicated would only cause desertion of the nest by the female. Thereupon, Keller and associates attributed approximately 70 per cent of the nest destruction to badgers, and completely eliminated ground squirrels as a source of destruction of sage chicken nests. Their conclusions on the nature of ground squirrel predation are contrary to the beliefs of this worker, although their observations on badger predation on sage chicken nests agree closely with those made by this writer.

In Wyoming, badgers accounted for 36 per cent of the nest destruction. The nests were generally dug out, and the nest lining and egg remains widely scattered for several feet in the vicinity of the nest. Shells were badly crushed, and often buried in fresh diggings near the nest site. Normally, large areas were dug adjacent to the nest, and frequently hair samples were deposited on the brush over the nest. Tracks were often observed in the freshly-dug holes around the nest site.

Magpies were responsible for 11 per cent of all nests destroyed. The eggs customarily were rolled from the nest and holes from .5 to 1.5 inches in diameter pecked in the sides. Openings in the shells usually were clean-cut, and the shells were seldom crushed. Coyotes were of minor importance in nest destruction on the Dry Sandy—Pacific Creek plots.

Other potential predators on the nesting areas were skunks, weasels, mink, domestic dogs and cats, bobcats, and resident birds of prey. None of these species assumed much importance as destroyers of sage grouse nests, probably due to their relative scarcity on the nesting areas.

In so far as could be determined, the nesting census caused little or no disturbance to sage grouse nests, nor did it create conditions which might have contributed to subsequent desertion or destruction of nests by predators. Ground cover was very sparse or else completely lacking between sage bushes at the period of the nest census. Therefore, no prominent trails adjacent to the nests were created for predators to follow. For comparative purposes, 400 acres were censused between May 30 and

June 3, 1949, which was after the peak of hatching. Thirty-eight nests were located during this period. Thirty-nine per cent had already hatched; 45 per cent had been destroyed; and 16 per cent were deserted. Excluding nests which were deserted due to human disturbance while the plots were being laid out, these nesting figures compare very closely with those obtained prior to the hatching peak on ninety-three nests. Thirty-three per cent of these had hatched; 47 per cent had been destroyed; and 20 per cent were deserted. These data were confirmed further by determining the percentages of nests destroyed and deserted prior to their being discovered, as compared with the percentages of nests destroyed and deserted subsequent to being found. Considering the overall nesting data in 1949 and 1950, only 12.5 per cent of the nests were deserted after location, whereas approximately 50 per cent of the total nests observed were either in a state of destruction or desertion when discovered.

Desertion occurred on a higher percentage of nests in Eden Valley than at Pacific Creek. The former plots actually were censused earlier than those on the Dry Sandy—Pacific Creek area; hence, it could logically be assumed that a certain percentage of the deserted nests on the latter plots had been destroyed prior to the census. Therefore, it was probable that the number of destroyed nests located on the Dry Sandy—Pacific Creek plots was in excess of that actually occurring; similarily, the percentage of deserted nests on the same area probably was represented as being too low.

Desertion may occur due to nest disturbances from a variety of causes during the process of egg-laying or early stages of incubation. On three separate instances during the nesting census hens were flushed from nests with incomplete clutches, and on the following day the nests were deserted. A truck frightened an incubating hen and then ran over her nest containing seven eggs, breaking two of them, and desertion occurred. Another nest located in a shallow drainage basin was deserted after heavy rains had partially flooded the nest site. Human disturbances were responsible for several hens deserting nests located off census plots during the 1949-50 nesting studies. On two occasions bands of sheep caused birds to flush and simultaneously to flip eggs out of their nests. Sheep subsequently stepped on these eggs, destroying them, and desertion occurred in both cases. No instances were noted of sheep breaking up sage chicken nests by stepping into them.

It was not possible to determine whether immature females were more inclined to desert nests than were adult birds, although there is a possibility of this occurring.

The location of a great majority of nests under sagebrush and other brushy cover would normally prevent their destruction due to the movements of livestock. In a few cases, nests were placed on open ground between shrubs and could have been destroyed by livestock activities, but no destruction of this nature was recorded. There was no indication that livestock is a serious factor in nest destruction, although nest desertion from livestock activities was of frequent occurrence under certain conditions. Desertion was most prevalent in the vicinity of sheep bedgrounds. Bands of from 2,000 to 3,000 sheep acted as serious disturbances to nesting activities, while milling on bedgrounds. Almost all nests found deserted in Eden Valley contained full clutches which, upon examination, were found to be either unincubated or else in the early stages of incubation. This indicated that desertion occurred in late April and early May. It was particularly significant that several thousand sheep began moving into this area en route to their summer ranges coincident with this period of nest desertion. Nest desertion seldom occurred after incubation was well underway. Thereafter, the normal movement of sheep bands during daily feeding or trailing activities was not considered to be a serious factor in nest desertion. Several birds were observed to be flushed from nests by sheep with no evidence of desertion.

While absent from their nests, hens exhibited great tolerance to minor disturbances of eggs and nests. No desertion or destruction was known to have occurred as a result of handling eggs or photographing nests.

Several nests were photographed at close range without the hens flushing. No blind was used in photographing nests. In some instances, incubating hens could be approached and photographed at distances ranging from 2 to 3 feet. Occasionally birds were actually touched before they would flush. One bird hissed and pecked violently at a human intruder, and had to be lifted bodily from the nest in order that her eggs might be counted. When the intruder had moved a few feet away, she returned immediately to her incubation duties. However, this behavior was not typical.

Renesting

Batterson and Morse (op. cit.) have suggested that older hens lay at earlier dates than do young hens, but offered no evidence in support of their theory. During both years of the Wyoming nesting study, approximately two-thirds of the nests on both areas brought forth young during the first two weeks of the hatching period. The remainder of the nests hatched from one to three weeks following the peak of the hatch. The average clutch size of these late nests in Eden Valley was 6.8 eggs, as compared with an average clutch of 7.8 eggs for early nests. On the Pacific Creek plots, late nests averaged 5 eggs per clutch, whereas the nests hatching earlier averaged 7.4 eggs. In none of the cases involving late nests were hens banded or otherwise identified positively as yearlings or renesters. However, smaller clutches and the presence of new nests adjacent to destroyed or deserted nests might possibly indicate renesting attempts. The large number of females observed in flocks after mid-May indicated that renesting was of limited occurrence, and involved only those grouse whose nests were destroyed or disturbed during egg-laying or extremely early incubation.

Nest Densities

The number of nests per 40-acre plot varied from zero to nine (Table 10). Plots adjacent to main irrigation canals and creeks generally exhibited higher nesting densities than plots 1 or 2 miles from open water. There were indications that birds returned to the same locations year after year for nesting. Two hens nest-trapped in 1949 were recaptured in 1950 on nests located less than 200 feet from their former nesting sites. Nests were frequently located in old nest cavities, as evidenced by pieces of bleached egg shells lying nearby. However, no nests were found in nest cavities known to have been used during the previous year. It was a common occurrence to find new nests within several feet of old nests.

TABLE 10

Frequency Distribution of Nests per 40-Acre Plot—1949

Number of 40-Acre Plots	Number of Nests Occurring on Plots
4	0
7	1
3	2
5	3
8	4
2	5
1	6
2	7
2	8
1	9

Groups of from two to six nests were commonly found. Maximum densities encountered were six nests on 5.4 acres. In another instance, five nests were found on 6 acres. On restricted areas, densities of one nest per acre were found. Other typical patterns were three nests within 100 feet of each other and two nests within 40 feet. Five different pairs of nests were within 130 feet of each other; seven pairs were within 200 feet; and six pairs were within 250 feet.

Nesting densities on the Eden Valley area averaged forty-two nests per square mile for the 1949-50 period. Although not generally recognized, significantly higher average nesting densities (as well as total numbers of birds) were encountered on the Dry Sandy—Pacific Creek area. On this latter area, fifty-nine nests per square mile were found during the same period. When the female population density is expressed on the basis of total numbers of hens flushed from nesting plots, the Eden Valley tract supported forty hens per square mile, whereas the Dry Sandy—Pacific Creek area carried a hen density of fifty-eight birds per square mile. The mountain foothills region provided habitat superior to that afforded sage grouse in Eden Valley, the irrigation system and alfalfa fields on the latter area, notwithstanding.

Nesting Cover

Ninety-two per cent of all nests found were placed under sagebrush. Rabbitbrush was occasionally utilized for nesting cover with greasewood and shadscale being used rarely. The average height of the cover over all nest sites was 14 inches (Table 11). Hens were inclined to nest in short sagebrush of medium density, such as is found on drier sites, in preference to the dense, tall brush found along watercourses and on moist areas. In this respect, moderate grazing by livestock would actually seem to improve the characteristics of sagebrush cover for nesting purposes. The individual nest site invariably provided means for a quick and unimpeded escape for the hen if she was flushed unexpectedly.

Eight nests were located on bare ground between sagebrush clumps with no cover of any type over the nest sites. Five of these nest sites were situated on a single 40-acre plot. This atypical nesting pattern seemed to be well-established for the particular group of hens concerned, since the remains of several old nests on the same plot were in similar locations. One of these nests was placed between scattered tumbleweeds (Russian thistle) in a sandy borrow pit barely 20 feet from the edge of a paved federal highway (U. S. Route 187—1 mile north of Farson). Hundreds of cars passed this nest daily and its clutch of eight eggs hatched successfully.

Fecundity

The average number of eggs per nest in 1949 based on eighty complete clutches was 7.26. The maximum clutch contained nine eggs, and was recorded five times. The minimum clutch held five eggs, and occurred four times. Hatching success based on 326 eggs was 92.1 per cent. Eighteen eggs were either infertile or else chilled during incubation. These represented 5.5 per cent of the total number of eggs in nests hatching successfully. Eight eggs (2.4 per cent) were fertile but failed to hatch due to

TABLE 11
*Frequency Distribution of Nests
in Relation to Height of Nesting Cover—1949-1950*

Number of Nests	Height of Nesting Cover (In.)	Number of Nests	Height of Nesting Cover (In.)
8	0	12	20
1	6	2	21
1	7	4	22
2	8	2	23
9	9	2	24
22	10	2	25
22	11	3	26
27	12	2	28
24	13	1	29
30	14	2	30
25	15	1	32
18	16	1	34
11	17	1	39
15	18	2	40
10	19		

poor development, loss from nest, breakage during incubation, or disturbance during hatching.

In 1950, the average number of eggs per nest based on seventy-four complete clutches was 7.53. One nest was found with thirteen eggs; otherwise, the maximum clutch was nine eggs. Hatching success based on 362 eggs was 91.4 per cent. Nineteen eggs, representing 5.3 per cent of the total, were either infertile or had been chilled during incubation. Twelve eggs (3.3 per cent) were fertile but had failed to develop to the hatching stage. As noted in Figure 7, the hatching and fertility values of sage grouse eggs were almost identical in 1949 and 1950.

Weather and its Effect upon Nesting and Brood Survival

As an aid in studying the effects of weather upon the viability and hatching success of sage grouse eggs during the 1949 and 1950 nesting seasons, daily precipitation and temperature records were compared during the laying and incubation periods from April 16 to May 19.

Meteorological data were obtained from the U. S. Weather

Bureau Station maintained by the U. S. Soil Conservation Service at Farson, Wyoming, elevation 6,591 feet. Minimum daily temperatures and daily precipitation expressed in depth of snowfall have been plotted from April 16 to May 19 for the two-year period (Figure 9). The 1949 incubation period was characterized by seasonal temperatures and an almost complete absence of cold, wet weather; although heavy, wet snowfalls of short duration did occur on May 5 and again on May 13, 1949. In 1950, the incubation season was characterized by a period of prolonged cold and snowy weather accompanied by abnormally low temperatures. Daily minimum temperatures were substantially lower than average, and there were fourteen continuous days of cold, wet weather with snow from April 27 to May 10. Daily minimum temperatures dropped to 15° F. or lower on five separate occasions, two of which occurred during the height of the egg-laying period. In the latter cases, temperatures fell to 8° F. on April 23 and 24 and to 11° F. on April 29. During the egg-laying period it was noted that, prior to departure from their nests, the females invariably covered their incomplete clutches with loose grass, sagebrush bark, and other insulating material. The minimum daily temperatures recorded at the nest or ground level were from 8 to 10 degrees higher than minimum daily air temperatures recorded simultaneously at a height of 6 feet above the ground level. The covering and insulating material over the eggs probably further increased actual egg temperatures above the corresponding air temperatures.

During the last week of incubation in 1949 (May 15-22), a total of 1.48 inches of rain fell on the Eden Valley area. On May 5 and again on May 12 heavy snow storms occurred, depositing 5 to 8 inches of wet snow over nesting areas. No nesting mortality occurred as a result of rain and snow storms during the incubation period.

The viability, fertility, and hatching success of sage grouse eggs for the two-year period are plotted in Figure 7. Hatching success in 1949 and 1950 was 92.1 per cent and 91.4 per cent, respectively. No nests under observation were deserted or failed to hatch as a result of low temperatures or snow conditions

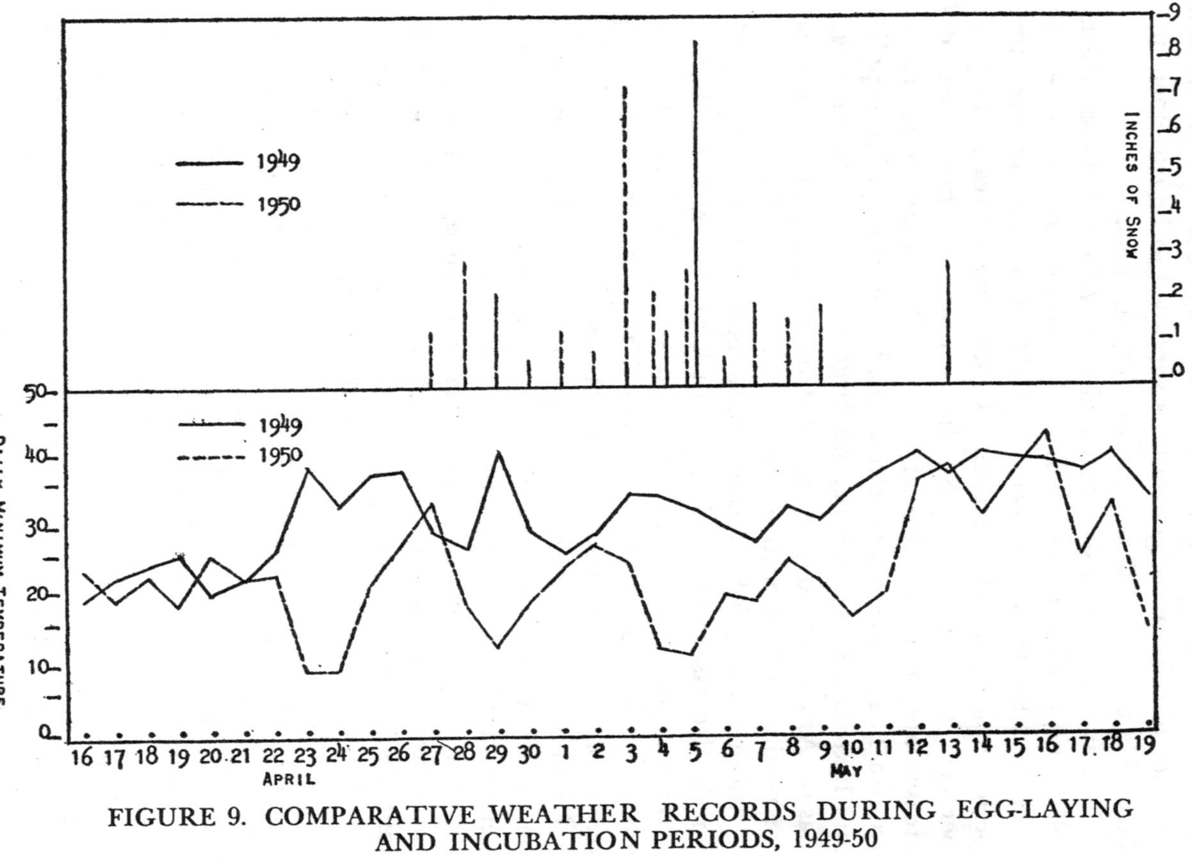

FIGURE 9. COMPARATIVE WEATHER RECORDS DURING EGG-LAYING AND INCUBATION PERIODS, 1949-50

which prevailed during the 1950 nesting season, even though weather of this nature generally is supposed to be extremely unfavorable to successful sage grouse nesting seasons.

The average clutch size increased from 7.26 in 1949 to 7.53 in 1950, and nesting success in Eden Valley increased from 44 per cent in 1949 to 61 per cent in 1950. However, in the same period nesting success on Pacific Creek plots decreased from 27 per cent to 23 per cent. Nesting densities on both tracts decreased approximately 10 per cent in 1950. Late spring runoff from a heavy winter snowfall created abundant water supplies and luxuriant forage throughout large areas of the Green River Basin. As a result, females selected more widely dispersed areas for nesting in 1950 than in 1949.

Another factor which could have been partially responsible for the decreased nesting densities in Eden Valley and adjoining areas was the removal of more than six hundred juvenile and adult females from ranches in Eden Valley during live-trapping operations in the summer and fall of 1949.

On both study tracts, the majority of nests hatched during the period from May 19 to June 7. The first week of this hatching period in 1949 was characterized by weather conditions ideally suited for brood survival. However, the entire region experienced severe, stormy weather from June 1 to June 7 with cold rain, sleet, snow, and strong wind storms occurring daily. These storms were accompanied by an average daily minimum temperature of 35.1° F. at Farson and 34° F. at South Pass. The weather at South Pass was typical of conditions existing on the Dry Sandy—Pacific Creek area. Farson is located in the center of the Eden Valley district. A total of .79 inches of precipitation fell at Farson and 1.41 inches at South Pass between June 1 and June 7. On June 2, several healthy-looking chicks were found dead within 25 feet of recently hatched nests. Juvenile mortality was high during this period due to several consecutive days of cold rain, sleet, and snow, accompanied by low temperatures.

In 1948, no serious juvenile mortality occurred as a result of unfavorable weather. From May 23 to June 7, average daily minimum temperatures were 7° F. higher at Farson and 2° F.

higher at South Pass than for the similar period in 1949. Precipitation at Farson and South Pass in 1948 was .39 inches and .93 inches, respectively, for this period. Furthermore, precipitation was of short duration with no cold wind storms.

Duration of the Nesting Period

The earliest nest was found on April 21, 1950, with three eggs. During the period from April 21 to April 29, 1950, eleven nests were located with partially completed clutches. These nests were not situated on regular nesting plots, and were closely observed to determine the rate of egg-laying, the incubation period, and the effects of intensive observation and disturbance upon nest survival. The chronology of egg-laying in ten of these nests is recorded in Table 12.

TABLE 12

Progression in Numbers of Eggs Laid in Ten Nests—1950

Date	Nest Number									
	1	2	3	4	5	6	7	8	9	10
Apr. 25	7	3								
26	8	4	5	3						
27		--	6	5						
28		5	--	--	3	3	8			
29		6	--	6	--	--	9	7	10	4
30		7	7	7	4	4		8	11	--
May 1			8	8	5	5			12	5
2					--	6			13	--
3					6	--				6
4					7	7				7
5										8
6										9

The rate of egg-laying based upon the history of twelve nests was one egg laid every 1.3 days.

Girard (op. cit.) as well as Rasmussen and Griner (op. cit.) have recorded the incubation period of sage grouse eggs at twenty-two days. Batterson and Morse (op. cit.) listed twenty-five days. Arbitrarily allowing one day following the completion of the clutch before incubation commenced, the length of the sage grouse incubation period, based on twelve nests studied in 1950 in Eden Valley, varied from twenty-five to twenty-seven days. This was considered to be from one to two days longer than the normal incubation period, and may have been prolonged due to a period of unusually cold and snowy weather which prevailed during late April and early May. In the case of ruffed grouse, Bump and associates (1947) have stated that under game farm conditions incubation normally requires twenty-four days. In the wild, however, development sometimes required several days longer; low temperatures or absence of the hen from the nest were listed as primary factors responsible for the longer period required for incubation in a natural environment.

The first sage grouse nest on the plots hatched on May 19. Allowing twenty-five days for incubation and nine days for completing a clutch of seven eggs, egg-laying would have commenced on or about April 15. The first mating attempts were noted on April 15. The 1950 hatching dates of fifty-five nests in the Eden Valley and Dry Sandy—Pacific Creek areas are plotted at weekly intervals on a percentage frequency basis in Figure 8. The large number of nests which hatched during the period from May 19 to June 1 indicated that females retire to nesting areas and commence egg-laying within a very few days after mating. Scott (1942) has stated that sage grouse lay no eggs for two weeks following the mating period, but this viewpoint seems at variance with observations made in Eden Valley. On April 16, 1950, a hen banded in Eden Valley during the 1949 nesting season appeared on a strutting ground 2 miles distant from her 1949 nest site. She was not observed to be mated on this date nor was she seen on this strutting ground on subsequent days. On May 22, 1950, the same bird (No. 968 plus colored bands) was captured on a nest less than 100 feet from her 1949 nest location. The 1950 nest had been discovered with three eggs on April 25. Egg-laying in this case had probably commenced about April 20, only four days after an appearance of this hen on a strutting ground. It is of interest to note that four other strutting grounds were located closer to her nest site than the one on which she was observed during the mating period.

The duration of hatching in Eden Valley extended until June

22, whereas at the higher Pacific Creek elevations hatching was not completed until July 1. Observations and brood studies conducted at altitudes ranging from 7,500 to 8,000 feet indicated that the hatching period is correspondingly delayed toward the lower limits of timber.

A hen which completed her clutch on April 30 was still incubating the same eggs, obviously chilled or infertile, when last observed on June 25.

The weekly hatching-groups of nests studied on the two tracts have been projected throughout the late summer and fall period to depict the percentage age composition of immature birds at successive monthly intervals during any hunting period selected (Figure 8).

Overgrazing and its Effect upon Nesting

As a result of overgrazing by sheep, a small parcel of land at the lower end of Eden Valley lost its attractiveness as a nesting area for sage grouse. Nesting plots of 200 acres in this sector produced ten and seven nests in 1949 and 1950, respectively. This summer range allotment has been overgrazed heavily by a local band of sheep for extended periods during the spring, summer, and fall months, resulting in the elimination of most of the herbaceous ground cover, with only sagebrush and other shrub cover remaining. The soil is very sandy and has blown severely, exposing the basal portion of the sagebrush to literally create hummocks of vegetation. The nesting density on this overgrazed tract was one nest per 23.5 acres, in contrast with a density of one nest to every 9 acres on adjoining nesting plots used moderately by sheep as they trailed to and from summer ranges in the mountains.

Excluding the incidence of nest desertion caused by migrant bands of sheep, there is no indication that moderate utilization of the range in the upper Green River Basin by sheep under the present grazing policies has created any conditions seriously detrimental to sage grouse occupancy and use. Some aspects of livestock utilization and game abundance in other parts of the sage grouse range, including a discussion of range use in former years, will be presented in a later section of this report.

Immature Stages

Physical Properties of Eggs

Weights and measurements of sage grouse eggs were obtained from clutches found in deserted nests (Table 13). The shape of the egg was obtained by taking length and width measurements and was based on 174 eggs representing all or part of thirty-three clutches. Weights were obtained on fifty-eight eggs from nine separate clutches in varying stages of development. In order to permit comparison with eggs of other gallinaceous birds, the weights and measurements of a pheasant clutch from Eden Valley, as well as extensive data on ruffed grouse eggs and chicks collected by the New York ruffed grouse investigations, have been included. The average live weight of newly hatched ruffed grouse chicks amounted to 72.4 per cent of the whole egg weight. A similar relationship between sage grouse chick and egg weights revealed that newly hatched chicks weigh 72.1 per cent of the average whole egg weight prior to hatching (Table 13).

Growth and Development

In common with other members of the grouse family, sage grouse chicks are precocial, being covered with feathers known as the natal down. Within a few hours after hatching, the chicks are completely dry, very active, and ready to leave the nest. Several nests were observed during the actual hatching stages. One nest, checked at 11:30 a.m. on June 12, had five partially opened eggs and two in the pipping stage. The following morning at 7:00 a.m. the female was flushed from this same nest and seven chicks scattered in all directions. After much diligence and repeated efforts these newly hatched chicks were reassembled in their nests and photographed. Another clutch of eight eggs was found to be slightly pipped at 6:00 p.m. on May 20. At 6:00 a.m. on May 22 the hen was still on this nest, but at 8:30 a.m. on the same date the hen and all chicks had departed. A third nest with seven eggs was observed in the early pipping stage at 6:00 p.m. on June 10. Observation of the nest at 10:00 a.m. on June 11 flushed the female, and revealed four newly hatched but inactive chicks and three well-pipped eggs. At 7:00 p.m. that evening the remaining eggs had hatched, and the hen and seven chicks had disappeared.

Girard (op. cit.) observed a brood leaving the nest site seventeen minutes after the last chick had hatched. The hen and

TABLE 13

Physical Properties of Sage Grouse Eggs and Newly Hatched Chicks as Compared with Pheasants and Ruffed Grouse

Sage Grouse Eggs (Eden Valley)	Number of Eggs	Average Value
Weight (All stages of development)	58	37.99 gm.
Weight (Early development)	18	40.48 gm.
Weight (Late development)	40	36.87 gm.
Shape:		
length	174	54.97 mm.
width	174	37.96 mm.
Pheasant Eggs (Eden Valley)		
Shape:		
length	12	44.2 mm.
width	12	34.3 mm.
Ruffed Grouse Eggs (New York)		
Weight	377	17.8 gm.
Shape:		
length	392	38.5 mm.
width	392	29.0 mm.
Newly Hatched Sage Grouse Chicks	Number of Chicks	Average Value
Live Weight	9	27.4 gm.
Newly Hatched Ruffed Grouse Chicks		
Live Weight	189	13.0 gm.

brood headed leisurely toward water that was 162 yards from the nest, requiring about two hours to travel this distance.

A hen and her brood of eight chicks in Eden Valley were observed within 100 yards of their nest site six weeks after the date of hatching. This bird had been trapped on the nest and leg-banded with colored plastic bands.

Young sage grouse develop rapidly and at the age of from one to two weeks are capable of making short flights of several dozen yards. During the first two months of life, most of their current energy is expended in a single flight. Thereafter, they are able to fly several hundred yards when flushed, and may make several consecutive flights.

Girard conducted flight experiments with about two dozen chicks between five and six weeks old. These birds averaged 163 yards in a single flight at an average height of 30 feet and at a speed of almost 30 miles per hour.

Weights of juvenile and adult sage grouse trapped in Eden Valley were obtained at approximately weekly intervals commencing July 7, 1950, and continuing until November 1, 1950. As far as possible these weights were taken weekly from a sample of birds representing the two sex and age classes handled during live-trapping operations. The weights of 155 mature males, 129 juvenile males, 88 mature females, and 116 juvenile females have been plotted as growth curves (Figure 10) to show the average weekly increase in weight of juveniles as they approached the weight levels of mature birds. The weights of mature males and females remained fairly constant throughout the summer and early fall period. Mature males averaged 71.6 oz. (4 lb. 7.6 oz.) throughout this period, whereas mature females averaged 41.3 oz. (2 lb. 9.3 oz.) in weight. By September, juvenile males had attained slightly more than 60 per cent of the weight of mature males for the same date. In gross appearance they seemed to be nearly as large as adult males; actually, on the basis of weight, they were less than two-thirds grown. During the same period juvenile females had reached a 73 per cent weight level, as compared with mature females and were less than three-fourths grown. By September 1, the weights of juvenile males were

Nests are rarely built on open ground with no cover. Typical clutches average seven eggs

Common nest site between two sagebrush clumps

Nesting activities in Eden Valley commence about April 20. Maximum clutches generally contain nine eggs

Clutches with five eggs usually indicate a renesting attempt

Grouse nests are frequently placed under a sagebrush limb

Destruction by ground squirrels accounted for more than 40 per cent of all nests found destroyed

Badgers caused a 36 per cent nesting loss on the Eden Valley and Dry Sandy—Pacific Creek plots

Rabbitbrush is occasionally utilized as nesting cover

Pintail duck nest situated on a sage grouse nesting plot more than one mile from water

Sage grouse nest and eggs after a May snowstorm

No nesting mortality occurred during the study from cold weather and snowstorms

An occasional hen displays no fear of humans

Nest desertion was improbable during the late stages of incubation

Ninety-two per cent of all nests observed were located under sagebrush

Nests are frequently well-concealed in dense sagebrush clumps

Incubating hens are well-camouflaged regardless of the nest site

Sage grouse incubation period varied from 25 to 27 days in Eden Valley

Average nesting densities varied from forty-two nests per square mile on Eden Valley plots to fifty-nine nests per square mile on the Dry Sandy—Pacific Creek tract

The nature of sage grouse courtship and breeding behavior prevents repeated nesting attempts throughout the spring and summer months

A ruffled sage hen defends her eggs and newly hatched chick. Majority of nests hatch between May 19 and June 1 in Eden Valley

Newly hatched grouse chicks and nest site amidst sagebrush, grass, and prickly pear cactus (*Opuntia*)

Large "clocker" or incubation dropping adjacent to a nest of hatched eggs

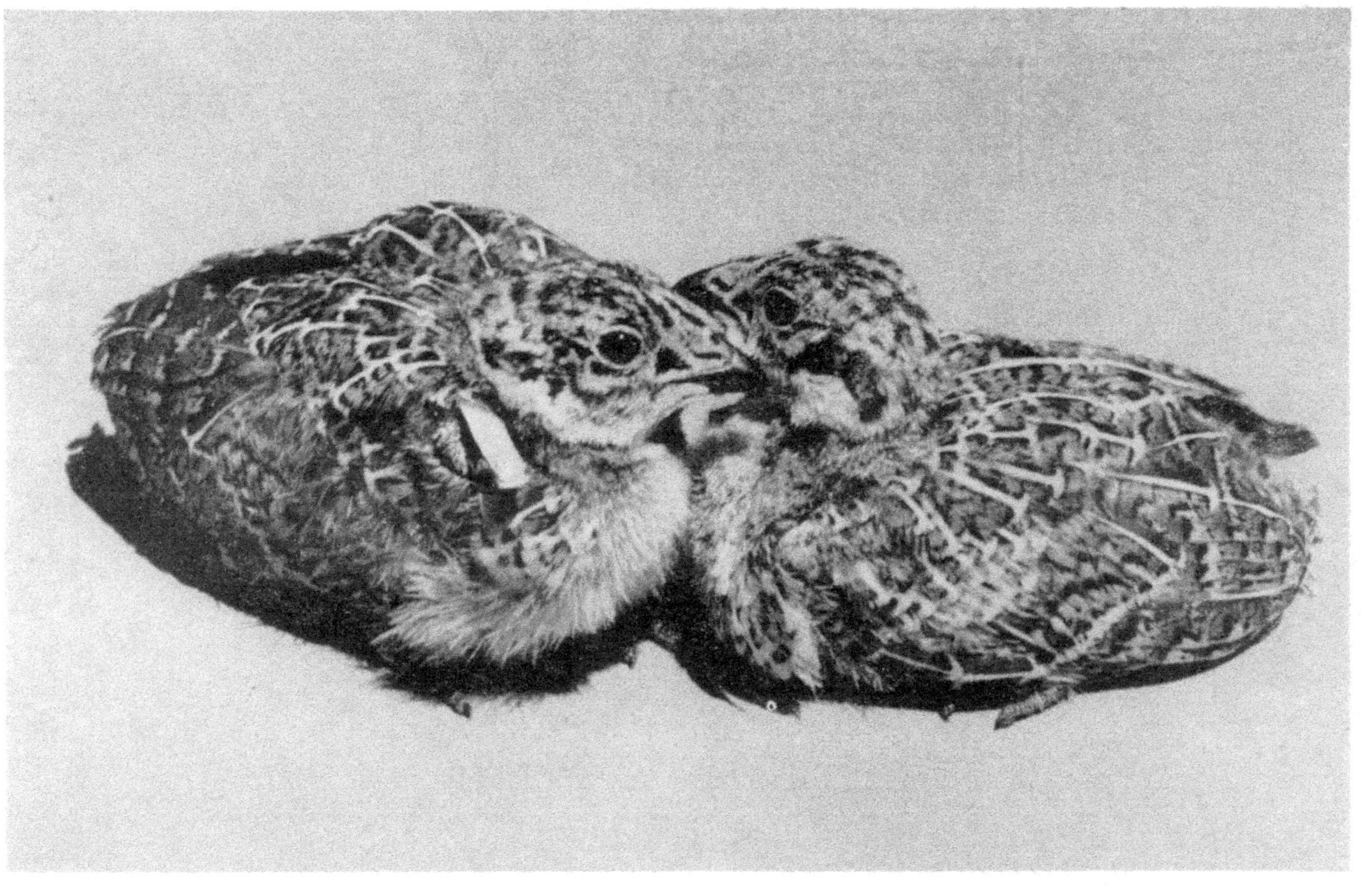

Ten-day old grouse chicks showing aluminum wing clip used in marking young birds

Day-old sage grouse chicks

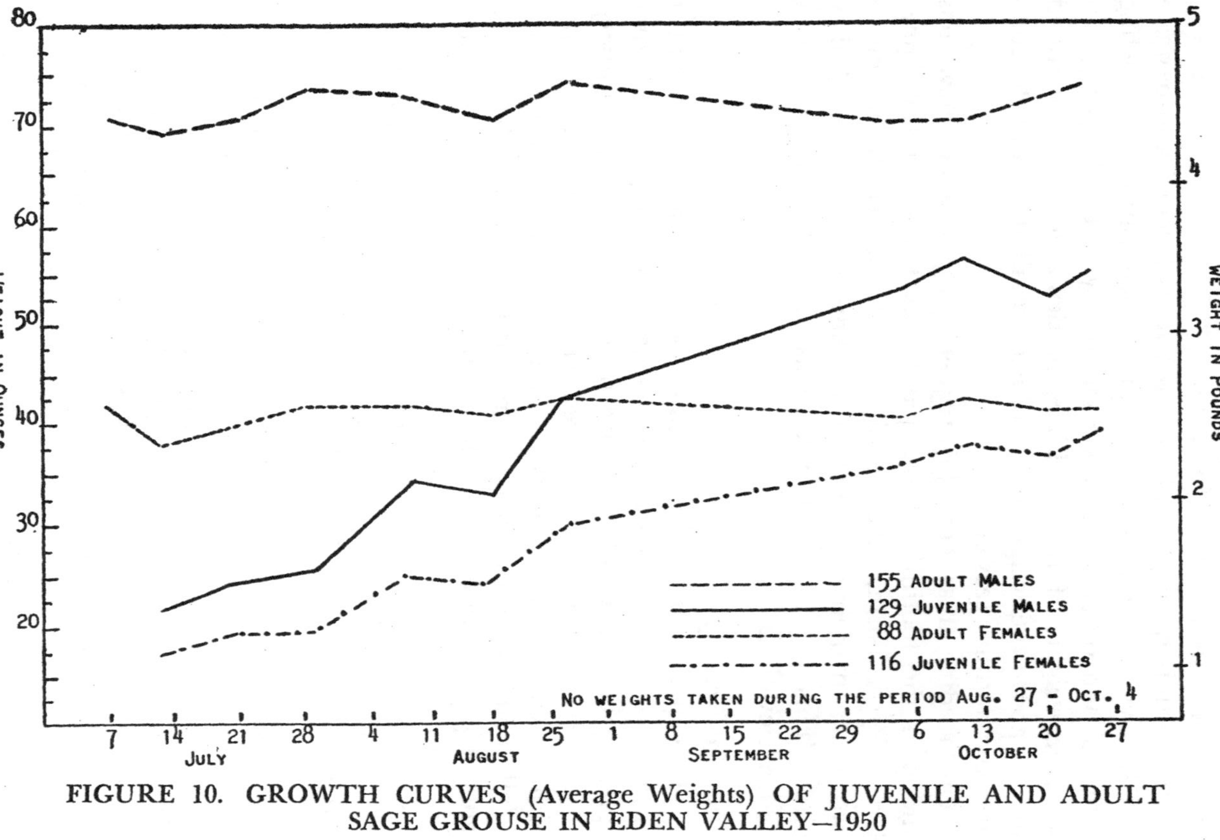

FIGURE 10. GROWTH CURVES (Average Weights) OF JUVENILE AND ADULT SAGE GROUSE IN EDEN VALLEY—1950

practically the same as the weights of mature females. Comparative measurements of adult and juvenile birds handled during October are presented in Table 14.

In November, the juveniles of both sexes reached approximate weight levels of adult birds. With the acquisition of breeding plumage and other structures associated with the breeding season, the weights of mature birds, particularly males, increased sharply above the weight levels which prevailed during the summer and autumn seasons.

On breeding areas located in regions lying at higher elevations and near the foothills of the mountains bordering the upper Green River Basin, juvenile birds were retarded from one to two weeks in their development as compared with Eden Valley birds. This was due to the later hatching dates prevailing at higher elevations.

Plumage Development

The sequence of the feather moult and plumage development in sage grouse was not studied in detail. Although many other upland game birds and waterfowl have been reared in captivity, little success has been experienced with sage grouse. It was not within the scope of this study to attempt to rear sage grouse artificially and such experiments were not conducted. The inhabitants of practically every ranch within the range of sage grouse have probably at one time or another slipped a few sage grouse eggs under a domestic hen for reasons of curiosity. I have heard of many such occurrences, and know of several cases of successful hatching, but have no records of chicks surviving for any length of time.

Since sage grouse chicks of known ages were extremely difficult to obtain under field conditions, plumage studies and age identification at weekly intervals based upon feather development were not considered feasible in this study. Sufficient observations were made on young chicks and juvenile birds handled during the course of live-trapping operations to indicate that the sequence of feather development in sage grouse closely approximated that encountered in other grouse species, particularly as outlined for the ruffed grouse by Bump et al. (op. cit.).

The chicks are in the natal down stage at hatching, and are strikingly marked. The general feather pattern of grayish-black is blotched with patches of brown and black on the body, wings, and head. Tips of the juvenile primary and secondary feathers are visible at hatching. The tail at this stage of development consists of a tuft of downy feathers.

TABLE 14

Comparative Measurements of 82 Adult and 111 Juvenile Sage Grouse in October

	57 Adult Hens	57 Juvenile Hens
Average weight	2 lb. 9.3 oz.	2 lb. 4.4 oz.
Weight range	2 lb. 4 oz. - 2 lb. 14½ oz.	1 lb. 14 oz. - 2 lb. 12 oz.
Average head length	71.4 mm.	70.5 mm.
Length range	66 - 75 mm.	67 - 74 mm.
Average middle toe length	55 mm.	54.4 mm.
Length range	48 - 61 mm.	50 - 57 mm.
Average tarsus length	56.3 mm.	55.6 mm.
Length range	52 - 62 mm.	51 - 61 mm.
	25 Adult Cocks	54 Juvenile Cocks
Average weight	4 lb. 8.6 oz.	3 lb. 6.5 oz.
Weight range	4 lb. 0 oz. - 5 lb. 1 oz.	2 lb. 8 oz. - 4 lb. 1 oz.
Average head length	83.1 mm.	80.1 mm.
Length range	79 - 88 mm.	76 - 84 mm.
Average middle toe length	66 mm.	66 mm.
Length range	62 - 71 mm.	62 - 69 mm.
Average tarsus length	64.4 mm.	67 mm.
Length range	60 - 68 mm.	62 - 71 mm.

From six to eight weeks after hatching (June 20—July 1 in Eden Valley), young birds have essentially acquired their full juvenile plumage, although the wing and tail feathers are still growing. The natal down has almost completely disappeared and the birds closely resemble the adult hen in plumage characters. The body feathers, however, are soft, fluffy, and fit loosely on the body in contrast to the firm, close-fitting feathers of the hen. By mid-August, young birds have acquired their juvenile plumage, although the wing and tail feathers are not fully developed. The juvenile bird's tail is several inches shorter, and the individual feathers are not as sharply pointed as in the adult.

The juvenile tail and wing feathers are moulted within two or three weeks after attaining their full development. With the completion of the postjuvenile moult, the bird acquires its adult or first winter plumage. The moulting of juvenile flight and tail feathers is a gradual process, continuing throughout most of the summer period. By mid-October, the majority of the immature birds have attained their adult plumage. During the late summer and fall seasons it is extremely difficult for an observer to distinguish immature birds of either sex from the adult hen, since their weights and external appearances are similar.

Field observations, as well as a collection of study skins, revealed that immature male sage grouse began to acquire their first nuptial plumage in October, and further showed that this plumage was not attained completely until April or May of the following spring. Adult and immature males could be readily identified during this period due to variations in the character of their nuptial plumage and the differences in length of their tail feathers. Once male birds acquired their nuptial plumage, there was no trouble in separating the sexes in the field since a considerable portion of the stiff, white nuptial feathers on the breasts remained visible throughout the nonbreeding season. The tail feathers of fully grown immature males averaged more than 60 mm. shorter than the tails of adult males. Immature and adult females were indistinguishable in the field once the young birds had acquired their first winter plumage. Immature birds normally retained their two outer juvenile primary feathers until the acquisition of their second winter plumage. These juvenile primaries were ordinarily not moulted until August or September of the year following their hatching.

The nuptial plumage of the adult male is characterized by the striking zones of black and white feathers on the throat, neck, and breast region, and by the addition of a filigree of lyre-shaped filiplumes on the crown.

Adult birds of both sexes commenced their postnuptial moult at the completion of reproductive activities. In the case of males, this moult began upon the cessation of strutting, and with females upon the completion of nesting duties or rearing of the young. This moult usually commenced by early June and involved body or contour feathers first, and was ordinarily completed by late September when the birds had fully acquired their winter plumage. The wing and tail feathers were customarily moulted in July and August.

In the late summer and early fall, juvenile birds of both sexes exhibited a triangular patch of feathers showing dark spots and light vertical streaks on the upper breast. This patch of feathers was quite distinct from adjacent feathers and presented a disrupted pattern of plumage development on juvenile birds in contrast with the solid feather pattern over the entire neck and breast region of adult hens. By late October, immature males commenced to display the black throat and black neck band so characteristic of cocks in the nuptial plumage. By early December, two patches of white feathers began to show on the upper breasts of immature males in the region of the bare tracts overlying the air sacs. From this point the nuptial plumage became more prominent throughout the winter months, and reached its maximum development for the first breeding season by late April or May.

Age and Sex Identification of Immature Birds

In addition to the age characteristics based upon plumage differences previously listed for juvenile birds, other features may be useful to distinguish immature birds from adults. In common with at least some other members of the grouse family, notably ruffed grouse, sage grouse do not customarily moult their two outer primaries during the postjuvenile moult, but retain these primaries throughout their first winter. As far as could be determined these two outer primaries of juvenile origin were not lost until the first postnuptial moult in the late summer of their second year. As pointed out by other investigators, the tips of these two primaries conform to the pointed character of the juvenile flight feathers, in contrast to the more rounded tips of the adult primaries. The worn and frayed tips of these pointed outer primaries further indicate their juvenile character, especially in the winter and spring months.

A saclike pouch or diverticulum on the dorsal surface of the cloaca, commonly called the *bursa Fabricii,* is normally present in juvenile and immature birds. The validity of this character as an indicator of age in sage grouse was not extended beyond the fall season, although in other upland game birds it has proved of value in distinguishing birds of the year from older individuals.

Girard (op. cit.) has described the mandible test for separating young sage grouse from adults and considered it to be a reliable method until the birds are about eight months old. When grasped by the thumb and forefinger, the lower mandible of the young bird will ordinarily not support the weight of the body without bending. The lower mandible of an adult bird will normally meet this test without breaking. The mandible test has been used to advantage on pheasants at game checking stations during hunting seasons, and has proved to be reliable on sage grouse under similar circumstances.

The sternum of juvenile birds exhibits a soft, flexible, and cartilaginous tip, whereas the sternum of adult birds is completely ossified throughout its length, and is solid to the touch.

During the fall season the toes of young birds are colored a light yellowish-green. Toes of mature birds are dark green in the fall, although they become somewhat lighter in the breeding season.

During the early phases of the study determination of the sex of juvenile sage grouse appeared difficult, since no marked variation in plumage existed. However, sage grouse live-trapping operations afforded an opportunity for handling and observing large numbers of birds in confinement. At an early stage in their development variations in the size of individuals in a brood became apparent, since adult males are approximately twice as large as adult females. These variations in size were utilized to distinguish the sex of young birds. They were particularly useful in banding operations, since males and females require aluminum bands of different sizes on their legs.

Prior to the completion of the postjuvenile moult, sex determination of juvenile birds was accomplished by visual inspection of the differences in size and shape of the head and length of the toes. In both of these respects males were noticeably larger. For example, the middle toe of male birds will average a toe nail length longer than the middle toe of hens of the same age.

In order to establish the validity of sexing juvenile birds on the basis of size, a series of weights and measurements were taken of young birds trapped and sexed in October. A comparison of the live weights and the lengths of the heads, middle toes, and tarsi of fifty-seven juvenile hens and fifty-four juvenile cocks revealed that no overlapping of the measurement ranges existed between the sexes in any of these categories (Table 14). Except for occasional renesting attempts, the hatching dates normally embraced a relatively short period of time in late May and early June. Consequently, almost all broods hatched, developed, and reached maturity at approximately the same time. During the summer and early fall period there logically would be a degree of overlapping between the sexes in the measurements used above due to the slight disparity in hatching dates. However, with the experience gained from handling thousands of birds during live-trapping operations, it was possible to determine age and sex of immature sage grouse with a very high degree of accuracy. The large number of birds available at game checking stations would afford equal opportunities to develop accuracy and acquire experience in aging and sexing sage grouse, particularly since the gonads would be accessible for examination.

Parental Relations with the Young

Males assumed no role in nest construction, incubation, or rearing of the young. Except for occasional associations around watering places, or special feeding areas, such as alfalfa fields, there was little or no mixing of the hens and their young with mature cocks during the brood period. The first general assembly of the sexes and age groups following the courtship period occurred during the late summer and early fall seasons as the birds gathered in heterogeneous flocks of large size, preparatory to movement to their winter ranges.

Behavior Patterns of the Young

The juvenile plumage closely resembles that of the adult female and provides young birds with the pattern of protective coloration typical of hens. The plumage blends remarkably well with the variegated colors of the sagebrush as well as the brown aspects of the sage prairie soils. Young grouse are well-adapted for camouflage and survival. The art of quick escape or concealment is instinctive from the moment of hatching. Young birds respond to the hen's warning and alarm calls by dispersing on foot in all directions until they reach a safe hiding place. Their reaction to sudden disturbances is to squat down and to remain motionless, often in full view of the observer. Ordinarily, their power of flight is utilized when they are certain they have been discovered or when danger is close at hand. Regardless of the distance of flight following the approach of danger, the birds will invariably run several feet from the point of landing and hide again. Frequently birds a few weeks old will remain frozen in positions of supposed concealment in the face of human danger, and may be captured without attempting to fly. Adult birds exhibit similar behavior by relying upon protective coloration and lack of movement in order to escape notice by humans as well as predators. Flight seems to be a secondary method of escape for both young and adult birds, and is utilized only when birds are closely pressed or as danger becomes really imminent.

Hens perform a very important role in establishing the feeding habits of the brood. During the first few weeks of life, the selection of food items is governed by the chicks observing and imitating the feeding behavior of the hen, as she designates for them various plant and animal materials. Once the selective feeding pattern is acquired, the chicks drift away from the hen and feed independently of her guidance.

During periods of inactivity the hens spend considerable time brooding the chicks, at least until they reach ages of from four to five weeks. Observations further revealed that during late May and June, brooding of the chicks occurs frequently throughout the day.

Some hens are extremely intolerant to the presence of stray chicks from other broods in company with their own young. This is particularly true if the stray chicks are older than those in her own brood. On two occasions in June hens were observed to drive away young birds attempting to join their broods. Other hens apparently acted as foster mothers for isolated chicks and sometimes for several chicks from other broods.

With one exception, the maximum clutch encountered in this study contained nine eggs. Out of 217 broods observed during June, 1948, nine broods contained chicks in excess of the maximum egg clutch. The maximum number of chicks observed with one hen was thirty-two, varying in age from two to five weeks. This particular hen successfully brooded twenty-seven of these chicks at one time with only a few heads peering from underneath her body. Several chicks in this enlarged brood appeared sickly and trailed far behind the remainder of the brood. A rather loose family organization prevailed in those areas which supported large concentrations of broods. Several of these areas existed on the Dry Sandy—Pacific Creek region where favored nesting areas bordered marshy, wild hay meadows utilized by the birds for daily feeding and watering. An extensive denuded plot adjacent to the upper Hay Ranch meadows on Pacific Creek afforded excellent opportunities to observe brood and family organizations, since hens brought their chicks across this bare area from the sagebrush en route to the meadows. Several dozen broods could be watched in this manner in a single evening. The number of chicks in individual broods varied considerably as young birds shifted back and forth from one group to another while crossing this clearing.

As has been stated, chicks of an individual brood are often scattered widely from the hen at times of disturbance. In an attempt to evaluate the effects of this type of brood dispersal as a juvenile survival factor, observations were made upon family groups which were purposely scattered. In all cases the hens re-assembled their broods in rapid fashion, seldom requiring more than one hour to accomplish the regrouping. The hen custom-arily occupied a point of vantage, either on a slight rise in eleva-tion or on a sagebrush clump, and by repeated calls revealed her location to the chicks as they steadily made their way back to her. The chicks answered the hen with plaintive, two-syllabled whistle calls, audible to human ears at distances of several hun-dred feet. These high-pitched calls of the chicks were easily imitated by a human and, when used, would lure a hen with a scattered brood right up to the caller. This type of brood dis-persal was not considered to be a source of juvenile mortality.

Most hens defended their broods from human intruders and predators with vigor and determination. When danger threat-ened, the hens first warned their chicks to scatter; then attempted to lure intruders away by feigning wing injuries and flopping clumsily along the ground. Captured chicks produced calls of alarm, causing the hen to give battle by repeated flights at the intruder, accompanied by violent cackling and hissing. When flushed from their nests the majority of hens immediately flew for several hundred yards before alighting. On rare occasions a hen would walk reluctantly away from the nest site, only to re-turn in a few minutes and resume incubation.

Brood Studies

During the month of June, 1948, counts were obtained on 248 broods along car-census routes in the Dry Sandy—Pacific Creek area; 190 were judged to be complete brood counts. The census route was approximately 30 miles long and roughly circuitous. These brood observations were accomplished by two observers, and were based on ten censuses covering 262 miles. The maxi-mum number of broods observed on a single coverage of the census route was sixty-two in a distance of 27 miles. The average number of broods observed per mile of census route for June was .95, or approximately one brood per mile. During the period from June 9 to June 17, the average number of young observed per brood was 5.36, based on complete counts of fifty broods. From June 18 to June 30, the average number of young observed per brood was 5.00, obtained from complete counts of 140 broods. Mean number of young for all complete broods observed during June, 1948, was 5.09. The largest brood observed consisted of fourteen chicks, and several broods numbered from nine to twelve young. Clutches of over nine eggs were so rarely observed during the nesting census that hens with more than nine chicks were assumed to be brooding some chicks belonging to another hen.

Since the 1949 censuses were taken by only one observer, it was necessary, for comparative purposes, to correct the 1948 data accordingly. Therefore, the average number of broods observed per mile in 1948 was set at .47 per observer, or approximately one brood for every 2 miles. Similarily, 1.1 hens were seen per mile of census route.

In July and August, 1948, counts were obtained on 322 broods; 274 were complete brood counts. Brood observations for these two months were based on twelve censuses covering a total of 380 miles. The average number of broods observed per mile of census route was .42. The average number of young for all broods observed during July and August was 5.18, based on 274 brood records. The higher average brood count in these months may be explained by the larger size and increased activity of the older chicks, making them easier to observe. The average number of young per brood did not vary appreciably during the 1948 brood census, indicating that relatively little juvenile mortality occurred during the early brooding season. This was in contrast to a high nest mortality encountered during the nesting study.

During the 1948 brood census a total of 464 complete brood observations were made, embracing 2,389 young birds. Incomplete brood counts were obtained on 106 hens. In addition to the 570 females with young, 622 adult females without broods and 464 mature males were observed along the brood census routes. The percentage of hens with young amounted to 48 per cent of the total hens observed. This percentage figure compares reasonably well with the figure of 39 per cent which represented the percentage of nests hatching successfully from thirty-three nests studied at Pacific Springs in 1948.

The postbreeding season movements of adult birds to higher elevations may introduce a slight degree of error in determining nesting success from roadside brood and adult census figures.

For the entire brood period the production of young per female was 2.4 based on 1,192 females of all categories and 2,930 young. It was assumed for this determination that hens observed with incomplete broods actually had the average number of chicks. The unbalanced sex ratio which existed among adult birds on the censused area may be explained as a result of large numbers of hens moving into favored areas for nesting purposes and the postbreeding movements of adult cocks to higher elevations.

During late July and early August, 1948, the number of broods and adult birds encountered along census routes dropped appreciably. The warm, dry weather of mid-summer had by this time eliminated many sources of open water. It seemed that a major movement of sage grouse took place at this time. Observations indicated that the direction of this movement was to higher elevations and to areas supplying more abundant water supplies.

Numerous broods, unaccompanied by hens, were observed in early August. On several occasions broods containing from twenty-five to thirty-five young, 10 to 12 weeks old, were seen with no hens in the vicinity. Due to the large incidence of such occurrences, it was apparent that family units broke up as the juveniles approached the three-month age level. The young of these broods thereupon organized as a flock.

In 1949, field headquarters for the sage grouse studies were moved from the Pacific Springs area to Eden Valley. Participation in live-trapping and transplanting activities in Eden Valley in the summer of 1949 interfered somewhat with the brood censuses on the Dry Sandy—Pacific Creek area, although it was possible to make a few censuses at irregular periods throughout June, July, and August.

During June, 1949, counts were obtained on twenty-five broods along the census routes. Thirteen brood observations were actually complete brood counts. The average number of broods observed per mile of census route was .15, or one brood for every 6.4 miles. The average number of young for each complete brood was 4.74. Brood observations were based on six censuses covering a total of 160 miles. The average number of young for each incomplete brood observed was 1.66. The 1949 brood studies indicated that young birds on the study areas sustained a marked reduction in numbers, as compared with 1948. Unfavorable weather immediately following the hatching period was partially responsible for the lowered numbers of young encountered in 1949 (see discussion of weather and its effect upon nesting and brood survival).

Considering only the months of June, 1948 and 1949, there was less than a 10 per cent reduction in the average number of young per brood in the latter year. This might indicate that the chicks which hatched in May, prior to the onset of the unfavorable weather previously discussed, were better adapted for survival as a result of their greater age. The average number of hens per mile in 1948 and 1949 was 1.1 and .9, respectively, indicating that on the Dry Sandy—Pacific Creek study area an 18 per cent reduction in numbers of hens apparently occurred in 1949. Nesting studies indicated that a 12 per cent reduction in nesting densities occurred in 1950 below that encountered in 1949 on the Dry Sandy—Pacific Creek plots. None of these figures are necessarily indicative of a general downward trend in numbers over large areas of the birds' range, since nesting densities on local areas are closely related to forage and water conditions.

During July and August, 1949, counts were obtained on eighty-eight broods. Brood observations for these months were based on fourteen censuses covering 344 miles. The average number of young for each complete brood was 3.89. The average number of broods observed per mile of census route was .26, or one brood for every 4.6 miles.

In summarizing the 1949 brood censuses, it may be stated that eighty-eight complete brood observations were made, embracing 350 young birds. In addition to the 113 hens observed with young, 396 adult hens were counted along the census routes. Twenty-two per cent of all hens observed along the census routes were accompanied by young. This compares with 27 per cent of the hens which successfully hatched young on the area. Including hens in all categories, one was seen each mile of the census route.

Factors Causing Juvenile Mortality

Within a period of a few days or of one to two weeks, the element of weather is capable of exerting great influence upon the survival of juvenile sage grouse. However, when considered over much longer intervals of time, environmental influences such as predation, highway mortality, and illegal hunting exact heavier annual losses of young birds than does unfavorable weather.

Hail storms occur frequently in parts of Wyoming, and may at times kill young birds over local areas. Weather apparently exerts no serious effects upon sage grouse once they have passed the vulnerable early stages of their growth.

The loose family organization, previously discussed, is another factor contributing to juvenile mortality. Straying of chicks is most prevalent around feeding and watering areas supporting concentrations of birds. Heaviest concentrations are found around wild hay meadows in foothill areas and near alfalfa and clover fields in irrigation districts. Deaths from exposure, disease, and predation seem to account for most of the wandering chicks. Occasionally these lost chicks—and at times entire lost broods—join other family groups as evidenced by several families containing substantially more chicks than would have been expected from a knowledge of clutch sizes.

Drowning of chicks is normally not encountered with sage grouse, due to the low precipitation and good soil drainage prevailing throughout the birds' natural range. In the vicinity of irrigation systems, some drowning of young birds does occur though.

The nature of vegetative cover and the low fire hazard in the early part of the rearing season reduce the likelihood of prairie fires contributing to juvenile losses.

Since the habitat of sage grouse is essentially composed of

climax plants and mature land forms—representing the ultimate in ecological and physiographic development under the conditions of soil and climate now prevalent over the birds' range—environmental deficiencies in the form of food, cover, and water are believed to be practically non-existent as sage grouse decimating factors, once the breeding cycle has been inaugurated.

Contrary to the viewpoint expressed by Girard (op. cit.), it appeared obvious to this writer that sage grouse chicks do not require water within a few hours after hatching. Although nesting densities were considerably heavier in the vicinity of streams and ponds, many nests and broods were found several miles from any type of open water. Oregon reported that a brood was raised in captivity to an age of six weeks without water. Moisture in the form of dew may supplement other sources of water.

Losses to natural enemies probably constitute the greatest source of juvenile mortality. The mechanics of game populations in equilibrium with their environment indicate that a large percentage of the game birds hatched every year fails to survive the first year of existence. Sage grouse possess a significantly lower breeding potential, and hence are probably relatively longer-lived than birds such as pheasants, quail, and other grouse of woodland and prairie. No records of average longevity or population turnovers of sage grouse are available. Extensive live-trapping and banding activities conducted on this study should eventually furnish some answers to these problems. The pattern of weekly visitation exhibited by mature and immature cocks on four strutting grounds closely observed at Farson during 1950, afforded an index to the annual population increment supplied by immature birds. During the early phases of strutting activity in March only adult cocks were present on the courtship grounds. Coincident with the arrival of hens on the grounds and the onset of the mating activity in early April, immature cocks began to appear with regularity. From this point the curves reflecting census counts of males utilizing these strutting grounds increased noticeably until a peak of visitation was reached just prior to the cessation of seasonal activities (Figure 11). This increase in cocks represented immature birds, and amounted to approximately 20 per cent of the maximum number of males present on these four strutting grounds at the peak of strutting activity in early May. There was little variation in the total numbers of male birds utilizing strutting grounds in Eden Valley in 1949 and 1950, indicating that the population was not fluctuating very widely. These inconclusive studies would indicate an average longevity of approximately five years for sage grouse in Eden Valley.

Strutting ground and nest censuses conducted in 1949 and 1950 indicate that there was a spring breeding population of about 10,000 sage grouse in Eden Valley each year, three-fourths of which were females. Assuming that all these females hatched and raised the average clutch of 7.4 chicks, the potential fall population would have approached 55,500 young birds. Nesting losses for each of the two years initially accounted for a 48 per cent reduction in productivity. Approximately 10 per cent of the eggs in successful nests failed to hatch due to infertility or improper development. An adult to juvenile ratio of 1 : 1.33 was obtained from several thousand birds checked during the hunting season (August 27-September 10, 1950). Since trapping and transplanting operations had reduced the fall adult population by approximately 1,000 birds, the September 1 juvenile enumeration would have approximated 12,000 birds, out of some 26,000 young birds actually hatched. Juvenile mortality for the first four months of life had already accounted for 80 per cent of the original reproductive potential (number of eggs actually laid). Juvenile mortality could have theoretically accounted for 95 per cent of the average reproductive potential by the next breeding season, thereby leaving slightly over 2,000 immature birds to replace the 20 per cent adult losses. The net result the second year would be neither a loss nor a gain in numbers—a situation characteristic of a static population. Under the above conditions, if the fall juvenile population (12,000 birds) was to survive until the following breeding season, the total population of mature birds would have more than doubled, giving over twice as many birds the second year, although 80 per cent of the young produced in that particular season had succumbed to either nest or juvenile mortality.

Under average conditions the high magnitude of juvenile losses attributed to predation, disease, and accidents has little significance as far as maintenance of numbers is concerned. Profound and permanent changes created in major habitats as a result of altering land-use and ecological patterns have real significance in the perpetuation of sage grouse. Management directed toward preserving and improving major habitat areas combined with a maximum yearly harvest by hunting will prove far more productive of results—and will return much more on investment—than management stressing predator and disease control, rearing and transplanting of birds, and continually closed seasons.

Wherever transportation and agricultural systems have invaded sage grouse range, the species is forced to compete with man for space, while simultaneously sustaining mortality from accidents caused by man's structures and improvements. Birds are killed by cars on roadways and by flying into obstructions such as fences, telephone lines, and power installations. They are lured into hay fields and irrigation districts which are thrust into the midst of their range, where they are decimated by mowing operations. In these artificially created areas of concentration, disease and parasitism assume their greatest role in juvenile mortality.

There are no indications that either young or old birds are susceptible to poisoning from plants consumed in their natural diet. The toxic effects of insecticides on sage grouse in Wyoming have already been discussed.

Reports of finding dead sage grouse have been circulated following rodent control activities. Griner (op. cit.) presented some results of experimental studies by Rasmussen which indicated that sage grouse may develop a type of poisoning similar to botulism. This was presumed to have been caused by the birds consuming fly larvae produced on the carcasses of small mammals poisoned during rodent control. Field observations both by Girard and Griner in areas recently poisoned for ground squirrels revealed no actual cases of poisoning sage grouse. Experimental studies by Ward et al. (op. cit.) further indicated that there is slight danger of sage grouse being killed with strychnine-poisoned grain used in ground squirrel control operations.

Family Disintegration

Schwartz (op. cit.) stated that pinnated grouse chicks seldom are to be found with the hen after they are from eight to ten weeks old, and from this time on the family organization gradually disintegrates. Lehmann (op. cit.) had previously reported that young Attwater's prairie chickens leave their family groups at from six to eight weeks of age, although many young birds remain with the hen well into autumn.

Sage grouse brood studies conducted on the Dry Sandy—Pacific Creek area revealed that only a relatively small percentage of sage grouse chicks are on their own at ages reported for prairie chickens. Twenty per cent of 103 broods of sage grouse chicks observed from July 22 to 25 were without hens. At this period these particular broods would have varied from seven to nine weeks of age. On July 23 a group of twenty-nine young birds was seen with no hen present; and again on July 25, in a different location, thirty-five young birds accompanied by one hen were counted. Forty-one per cent of eighty broods observed from August 1 to August 15 were existing independently of hens, and six separate groups of young birds contained from fourteen to thirty-one individuals. On the basis of these data, it would appear that the majority of sage grouse family units had disintegrated by September 1. The high percentage of broods observed without hens in mid-August would indicate that young birds attain a stage of independent survival at ages from ten to twelve weeks, or at a period corresponding closely with the complete development of the juvenile plumage. Referring to the frequency percentage hatching figures for the Dry Sandy—Pacific Creek tract (Figure 8), approximately 50 per cent of the young birds would be from eleven to twelve weeks of age by mid-August.

Dates of Achieving Maturity

With respect to weight, size, and plumage development, young females achieved maturity in the fall of their first year, usually by mid-November. Sexual maturity was attained by the onset of the breeding season the year following their hatching. All evidence indicated that yearling hens performed nesting duties.

Young males achieved the approximate size limits of adults at a period corresponding closely with that of young females (November). In the early winter months (December and January), immature cocks began to undergo marked weight and plumage changes typical of the breeding period. At this period the young males commenced to exhibit a few white feathers in the throat region. Their weights of about 4½ pounds were approximately 1 pound heavier than that of mature hens and 1 pound lighter than mature cocks for this period. Their gonads were undeveloped and significantly smaller than the atrophied testes of mature males in the nonbreeding condition. During successive months throughout the winter and early spring the white feathers on the throat and neck regions of immature males became more prominent. Their weights remained fairly constant, and the testes showed no appreciable enlargement during the period from December through March. By mid-April the plumage of young cocks began to resemble the full nuptial plumage of the mature males. At this time the testes were approximately the same size as those of mature cocks, and the young cocks were then considered to be sexually mature. However, with respect to full breeding weight and nuptial plumage development, young males (yearlings) did not attain the features characteristic of fully-matured cocks until their second breeding season.

For comparative purposes, measurements of testes of adult and immature cocks were taken at intervals of approximately one month from December until June (Table 15). Microscopic examination of tissue from the testis of a yearling cock taken on April 23 revealed that spermatogenesis was occurring at that time.

It was probable that young males from late broods of the previous year did not become sexually mature until somewhat later in the season (May). Extensive observations of mating activity on the strutting grounds revealed that these yearling cocks, although in most respects sexually mature, took no active part in either the courtship activity or mating performances during their first breeding season.

TABLE 15

*Comparative Measurements of Testes of Adult and Immature
Sage Grouse from December until June*

Adult Cocks	Immature Cocks
Dec. 11–10 x 3 mm.	Dec. 13– 8 x 1.5 mm.
Mar. 19–21 x 12	Jan. 16– 7 x 3
Apr. 25–26 x 12	Feb. 4–10 x 2
May 1–27 x 13	Mar. 18–11 x 5
June 8–14 x 6	Apr. 19–23 x 8.5
June 9–13 x 7	Apr. 23–24 x 10

Courtship and Mating

Strutting Ground Censuses

Areas identical to those utilized for the nesting study were
used to determine the location, distribution, and occupancy of
strutting grounds. One area embraced the Eden Valley irrigation district (205 square miles), and the other included the Dry
Sandy—Pacific Creek territory (83 square miles) which extended
from Eden Valley to the foothills of the Wind River Mountains.
The latter territory extended 25 miles northeastward from the
rectangular-shaped irrigation district, and ranged from 2 to 7
miles in width. Elevations on these tracts varied from 6,400 to
7,500 feet. Except for the presence of a state highway and a
temporary oil drilling operation, the Dry Sandy—Pacific Creek
area was primitive sagebrush prairie.

During April and May, 1949, the lands on both of these tracts
were thoroughly reconnoitered for strutting grounds. With the
aid of binoculars and pickup truck this region was systematically
checked during the height of strutting activity each morning.
All strutting grounds on both tracts were believed to have been
located in 1949, and total counts were obtained on cock birds
utilizing them (Figures 3 and 4). No attempts were made to
enumerate females utilizing the grounds, although the seasonal
pattern of female activity was determined from day to day by
means of spot counts. Females exhibited a defined pattern of
visitation, although they were difficult to observe due to their
twilight appearances under poor light conditions, their small

size, and their inconspicuous coloration as they moved about in the brush.

Within the confines of the Eden Valley irrigation district twenty-eight separate strutting grounds were located in 1949. These grounds contained a total of 2,178 males, or an average of 78 males per strutting ground. The smallest number of males utilizing an individual ground was 14, whereas the maximum number was 400. Natural habitat for sage grouse was limited to approximately 167 square miles on the irrigation district. The remaining acreage was devoted to agricultural pursuits and irrigation systems. Density of the male population for this area was 13.0 birds per square mile. Strutting grounds averaged one for every 6 square miles of natural habitat.

On the adjacent, undeveloped area (Dry Sandy—Pacific Creek) sixteen strutting grounds were located in 1949. These contained 956 males, or an average of 60 males per strutting ground. The smallest number observed on any one ground was 6 males and the maximum number was 183. This entire area of 83 square miles furnished ideal habitat for sage grouse. The density averaged one strutting ground per 5.2 square miles.

The combined total on both areas was 3,134 males and forty-four strutting grounds on 250 square miles, giving an average density of 12.5 males per square mile and one strutting ground per 5.7 square miles. The number of cocks and location of strutting grounds on both tracts are listed in Tables 16 and 17 for the years 1949, 1950, and 1951.

Although covered less intensively, the same areas were checked in 1950 and 1951. Each of the grounds previously located in 1949 was utilized by birds in these succeeding years. In neither of the latter years were any newly created strutting grounds discovered. The number of cocks utilizing the Eden Valley grounds in 1950 was 2,316. When grouped by major locality within Eden Valley, the grounds carried almost equal male populations in 1949 and 1950 (Table 18). However, the 1951 strutting ground census revealed a 24 per cent decline in numbers of males below the 1950 figure. This reduction was believed to be largely due to the removal of several thousand adult and juvenile males

through transplanting operations as well as the hunting season
conducted in the late summer of 1950.

TABLE 16

*Sage Grouse Cock Censuses and Dates of Maximum Appearance
on Eden Valley Strutting Grounds—1949-1950-1951*

Strutting Ground No.* and Location	1949	1950	1951
1. West of Farson	April 11— 35	May 16— 77	April 12— 37
2. West of Farson	April 20—103	April 13— 99	April 27— 63
3. N.W. of Farson	April 22—212	May 5—220	April 12—114
4. Route 187	April 22—116	April 21— 92	April 12— 77
5. Haystack Butte	April 9— 64	April 21—100	May 9— 82
6. Big Sandy Dam	April 17— 14	April 4— 10	April 15— 2
7. Eden Reservoir	May 4— 29	April 25— 26	April 14— 23
8. Main Canal	April 18— 96	April 23— 89	May 5— 76
9. County Line	May 4— 31	April 25— 30	April 2— 24
10. Diversion Gates	May 20— 23	April 23— 36	May 5— 33
11. Dry Sandy	April 27— 17	April 22— 9	April 14— 6
12. Round Top Butte	April 25— 60	April 23— 46	April 14— 52
13. McMurray Homestead	May 7— 93	April 24—127	April 28— 85
14. Eden Canal	April 11— 63	April 15— 60	April 28— 45
15. Little Sandy Flat	May 7— 35	April 18— 42	May 4— 35
16. Little Sandy Flat	May 7— 18	April 18— 26	May 4— 24
17. Little Sandy Flat	May 7—119	April 20—149	April 28— 90
18. Little Sandy Flat	April 14— 79	April 22—118	April 28— 71
19. Wash'gton Draw	May 9— 50	May 16— 53	May 3— 36
20. Wash'gton Draw	May 9— 25	May 16— 75	May 3— 72
21. Wash'gton Draw	April 25—400	May 16—350	May 3—295
22. Wash'gton Draw	April 25—110	April 26— 85	May 3— 40
23. Tomich Ranch	May 7— 78	April 26— 90	April 13— 44
24. Eden Shelter Cabin	May 12— 51	May 4— 59	May 2— 52
25. Wash'gton Draw	April 23— 82	April 26— 70	May 2— 48
26. Wash'gton Draw	April 23— 56	May 1— 42	May 10— 10
27. Wash'gton Draw	April 24—101	May 1—108	April 13— 72
28. Trona Well	May 13— 18	May 4— 28	May 2— 19
Totals	2,178	2,316	1,627

*Numbers refer to strutting grounds shown in Figure 3.

TABLE 17

*Sage Grouse Cock Censuses and Dates of Maximum Appearance on
Dry Sandy—Pacific Creek Strutting Grounds—1949-1950-1951*

Strutting Ground No.[a] and Location	1949	1950	1951[b]
29. Dry Sandy Ridge	April 27—153	May 8—110	May 10— 55
30. Dry Sandy Ridge	April 27—107	May 8— 91	May 10— 57
31. Dry Sandy Dam	April 27—183	May 8—175	May 10—165
32. S.E. Dry Sandy Dam	May 3— 51	May 8— 61	May 10— 43
33. Middle Ranch Road	April 28— 15	May 9— 13	May 10— 9
34. Middle Ranch Road	April 28— 13	May 9— 11	May 10— 3
35. Rt. 28 Pond	April 28— 93	May 9— 91	
36. Rt. 28 Pond	April 28— 45	May 9— 49	
37. Superior Airstrip	April 29— 45	May 9— 34	May 14— 29
38. Superior Airstrip	April 28— 30	May 9— 25	
39. Superior Oil Well	April 30— 45	May 20— 42	May 18— 47
40. East of Superior Well	April 30— 6		
41. Oregon Trail Plot	April 30— 75	May 12— 75	May 14— 43
42. N. of Oregon Trail Plot	April 30— 30	May 20— 46	
43. N. of Rt. 28 Spring	April 30— 55	May 12— 60	
44. E. of Juel pens	April 30— 10		
45. Whitman-Hart Memorial	April 26— 65[c]		May 15— 42
Totals	956	883	520[b]

(a) Numbers refer to strutting grounds shown in Figure 4.
(b) 1951 census figures are incomplete.
(c) Off study area and not included in 1949 total.

On the Dry Sandy—Pacific Creek area all grounds located in
1949 were utilized in 1950 and 1951. No new grounds were discovered during the latter two years. Fourteen of the strutting
grounds used in 1949 were censused during 1950. The number
of cocks utilizing these fourteen grounds in 1950 was 883, as

TABLE 18

*Numbers of Strutting Cocks Grouped by Major Localities
in Eden Valley—1949-1950-1951*

Locality	1949	1950	1951
Little Colorado Desert	544	598	375
Eden Reservoir	256	236	214
Little Sandy Flat	882	1,000	753
Washington Draw	496	482	285
Totals	2,178	2,316	1,627

compared with 940 in 1949, representing a decrease of 6 per cent from the previous year. An extended period of unfavorable weather and a limitation on time available for censusing prevented making a complete enumeration of cocks on the strutting grounds on the Dry Sandy—Pacific Creek area in 1951, although several days were actually spent in an attempt to obtain reliable population figures.

Activity Patterns on Strutting Grounds

It was noted in 1949 that the number of cocks on most strutting grounds noticeably increased in late April and early May. Since Batterson and Morse (op. cit.) have stated that immature males and females appear on the strutting grounds later in the season than do the adult cocks, there was a probability that immature cocks were responsible for the increasing numbers of cocks seen on the strutting grounds in Eden Valley as the breeding season progressed. In order to facilitate recognition of immature males on the strutting grounds and to ascertain the time of year at which they reached sexual maturity, it was necessary to collect a series of immature and adult birds at monthly intervals prior to the breeding season. Commencing in December an immature and an adult male were collected and preserved as study skins, and significant data on gonad size, plumage development, and weight were recorded.

In an attempt to determine the causes for the increase in cocks encountered on Eden Valley strutting grounds the following procedures were instituted: (1) Four strutting grounds were selected for experimental censuses so that all of them could easily be visited by car and censused the same morning, thus assisting to determine whether or not males shifted from one ground to another on consecutive mornings. (2) A marking apparatus similar to that used by Moffitt (1942) on sage grouse in California was constructed and set up on one of the strutting grounds, so that several strutting males could be marked with colored analine dyes.

For the purpose of determining the reliability of strutting ground censuses as an index to populations, yearly trends, sex ratios, and breeding activities, the four experimental strutting grounds located adjacent to each other in a solid land unit just west of the Big Sandy Creek at Farson, Wyoming, were censused at regular intervals during the period from March 14 to June 4, 1950 and from February 25 to May 17, 1951. They were visited ninety-three times in 1950 and fifty-two times in 1951.

Prior to their appearance on regularly established strutting grounds, males engaged in strutting activities on their respective daily roosting areas as early as January. No males had commenced to use these grounds on March 5, 1950. When next censused on March 14, all four grounds contained males in active courtship displays. In 1951, males first moved onto these grounds on February 25 as a result of milder weather and a lack of snow cover. During both years an increasing number of males arrived on these grounds throughout March and early April. By mid-April their numbers had commenced to level off, and they remained fairly constant for the ensuing four weeks. Strutting activities on regular grounds lasted for about three months, and in Eden Valley usually had terminated completely by June. (Active strutting displays may be observed as late as mid-June in Jackson Hole, Wyoming).

The approximate weekly pattern of visitation exhibited by sage grouse cocks on these four experimental strutting grounds is plotted in Figure 11 for 1950 and in Figure 12 for 1951. Referring to the graphs it will be noted that on all four strutting grounds the numbers of cocks exhibited similar patterns of occurrence throughout the 1950 and 1951 strutting seasons. It should be emphasized that the trend of daily strutting ground

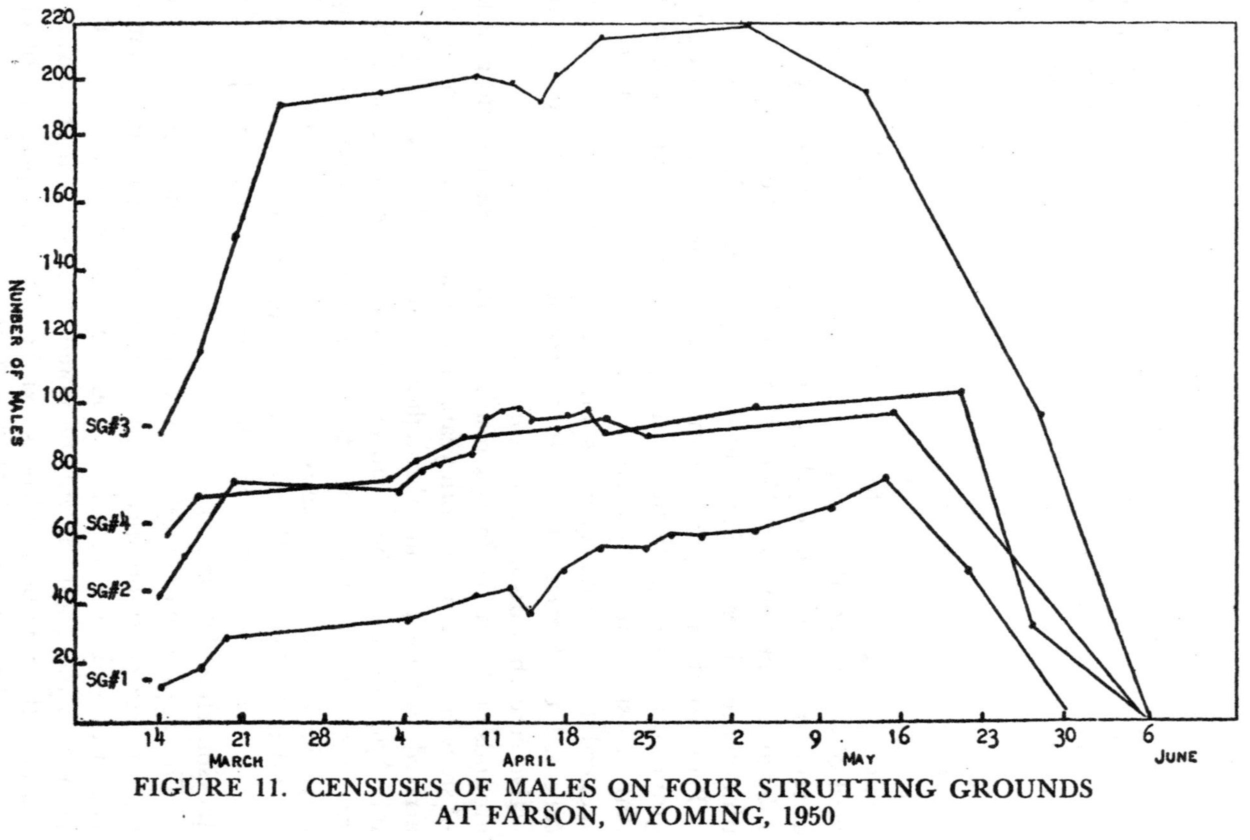

FIGURE 11. CENSUSES OF MALES ON FOUR STRUTTING GROUNDS AT FARSON, WYOMING, 1950

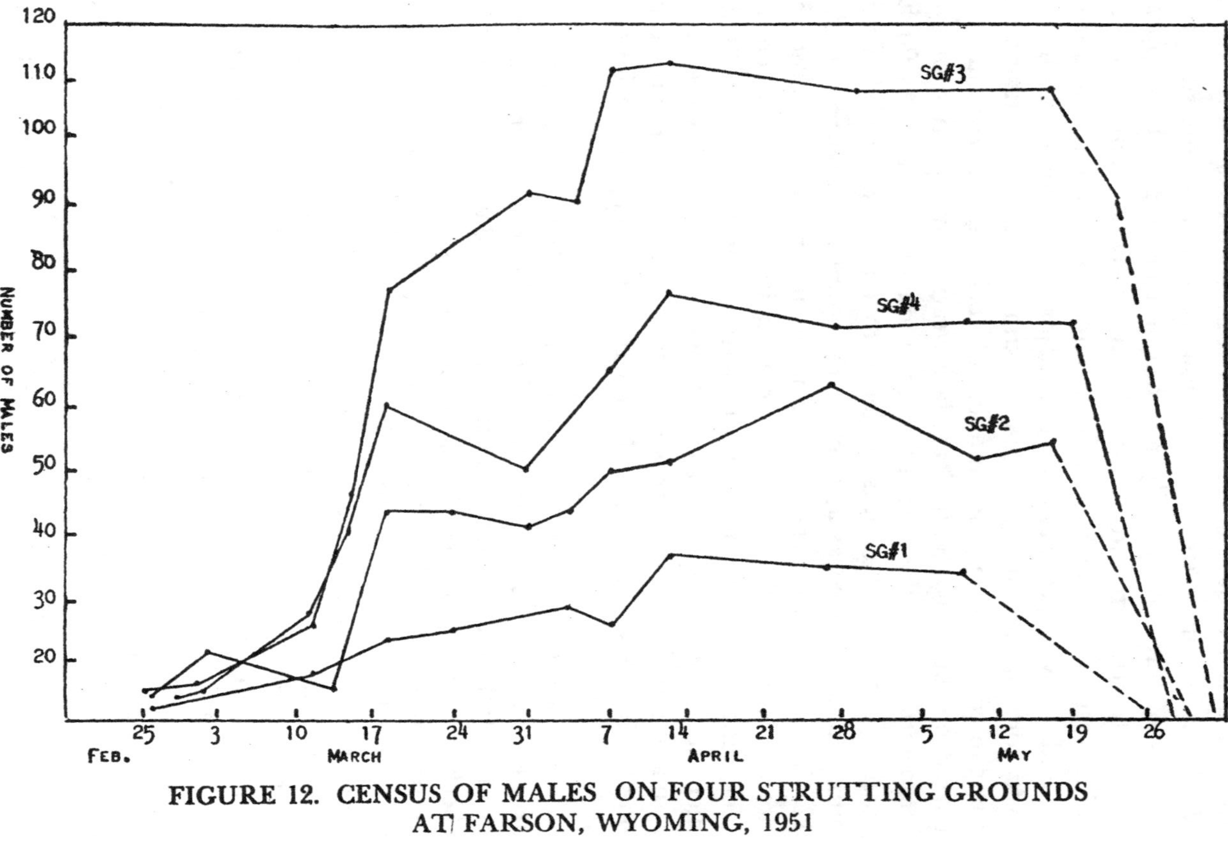

FIGURE 12. CENSUS OF MALES ON FOUR STRUTTING GROUNDS AT FARSON, WYOMING, 1951

visitation was dependent upon the occurrence of ideal weather and an absence of disturbance factors. On days with strong winds, rain, snow storms, and low barometeric pressure the activity of birds was greatly reduced and many failed to appear on the strutting grounds at all. On the many occasions in which golden eagles, prairie falcons, and buteos flew over the strutting grounds, it was customary for the majority of birds to flush and to fly for considerable distances. Maximum strutting activity, accompanied by the largest numbers of cocks on the grounds, occurred on mornings with clear, fair weather, irrespective of low temperatures and the amount of snow on the ground. As far as possible, only those counts believed to have represented maximum numbers of cocks appear on the graphs in Figures 11 and 12. The censuses on the four grounds under intensive observation in 1950 likewise showed a marked increase in numbers of cock birds following the peak of mating activity in April. This increase was not so apparent in the 1951 censuses, nor were there as many immature males on strutting grounds in 1951, probably due to the heavy hunting and large-scale trapping operations in 1950.

It was apparent from continuous daily observations that the arrangement and spacing of individual cocks on any one strutting ground follow a definite pattern day after day. As a further check on these observations, several cocks on one strutting ground were sprayed with colored analine dyes. Other cocks were identified as individuals by means of conspicuous feather patterns following the loss of certain tail feathers. All these marked birds were identified morning after morning occupying the same territory on the strutting ground. These individual territories were small in extent, and in a sense were defended against other males, although there was a great deal of overlapping of individual strutting areas among adjacent birds. There was no evidence that adult cocks shifted from one strutting ground to another on either successive mornings or at longer intervals. Prior to their selection of permanent strutting grounds immature cocks seemed to drift around promiscuously from one area to the other on consecutive mornings, particularly during the early weeks of the breeding period.

Efforts were made to determine the age classes of males utilizing the strutting grounds. The series of adult and juvenile study skins provided the basis for separating the mature and immature age classes encountered in the field during the spring. Adult males in full nuptial plumage exhibit long and finely pointed tail feathers. The black throat and black abdomen are in sharp contrast to an almost pure white breast region. Long, black filiplumes are present on the sides of the upper neck. In the courtship display these plumes are erected to form a lyre-shaped figure which adorns the crown. Adult cocks in the breeding condition average nearly 6 pounds in weight, with an occasional bird weighing up to 7 pounds. The increased weight of breeding cocks is caused by an enlargement of the lymphatic and vascular systems, as well as by an expansion of muscle and fat tissues in the air sacs regions. Still another factor contributing to an increased spring weight is the presence of layers of fatty tissue underneath the skin and in the body cavity, especially around the heart and the abdomen. Fatty tissue accumulates in the birds during the winter on a diet purely of sagebrush leaves. At the peak of daily strutting activity mature cocks in active courtship erect their tails in a semi-circular pattern, not unlike that of a strutting turkey gobbler. The air sacs and breast structures are distended almost to the ground.

In the late winter and early spring seasons immature cocks begin to acquire plumage typical of the adult birds. However, their tail feathers are noticeably shorter than those of adult males, and also are bluntly pointed and tipped with white. The breast region lacks the heavy, stiff, white feathers which adorn adult males. The filiplumes are not fully developed. Prior to mid-April the weights of immature males average about 4½ pounds, almost 2 pounds lighter than most mature males. During early phases of the strutting period (March) the immature males take no active part in the courtship displays, and seldom appear on the strutting ground proper. The few immature cocks that do appear, continuously move through the ranks of the strutting males, but never establish strutting territories nor engage in active courtship displays this early in the season. They are under constant harassment by mature cocks whenever they wander onto the center of the strutting ground. Other groups of immature males are frequently observed on the extreme periphery of the strutting ground and in many cases attempt to imitate the strutting performances. On April 4 nine immature males in one group attempted to perform the complete strutting act. Their tails were erected, and their air sacs were slightly inflated. The neck region was not greatly distended, and no audible plopping sounds were forthcoming. During this rather unique performance, an adult cock moved into their midst and proceeded to demonstrate the fine points of strutting to the rapidly maturing novices. Such was the pattern of immature activity during the early phases of the breeding season and essentially through the height of actual mating activities (or until about April 20).

By mid-April, groups of from four to six immature males are appearing regularly on the mating ground. In these instances they are not permitted to engage in any strutting activity, being chased vigorously by the dominant males whenever they approach mating spots. Following the peak of mating, immature males begin to move onto the main part of the strutting ground in considerable numbers, and commence to display strutting behavior characteristic of the older birds. In the absence of the hens, which have retired to their nesting assignments, there is considerably less competition and conflict between the young and old male age classes for strutting territories. The increased number of cocks which appeared in late April and early May on almost all strutting grounds under observation is a result of the presence of these now sexually-mature young males. Some of these birds had previously been present throughout the strut but were largely unnoticed due to their positions of concealment in the heavier sagebrush adjacent to the mating ground and to the absence of active courtship displays on their part. Others had irregularly visited the grounds, shifting around considerably in areas and company unfamiliar to them. Even though sexually mature, young cocks were not observed to mate.

Mating Behavior and Social Dominance

Scott (1942) presented the first detailed description of the mating behavior of the sage grouse. He has stated that 2 to 3 per cent of the male birds may be termed "heteroclite cocks" (sexually abnormal). According to Scott these heteroclites have about the size but not the full feathering of normal cocks. The form of the body is somewhat between the two sexes. They do not strut, and will not fight with other cocks. When one passes through the area, each stationed cock in turn chases him with such energy that the heteroclite frequently takes to flight. Occasionally one manages to sneak into a group of hens on a mating spot, and as he dodges around among them the master cocks frequently have some difficulty in driving him away. On one occasion Scott saw a heteroclite start to mate with a hen, only to be driven away by the master cock. In general, these birds behave more like hens than cocks. Their hormone system is evidently out of adjustment. Scott concluded by saying that further work was needed to explain the nature of heteroclites.

Scott's observations on his so-called heteroclites are essentially the same as my own observations of immature cocks on several strutting grounds in Eden Valley. In his work near Laramie, Wyoming, Scott made no distinction between immature and adult cocks; nor did he recognize the appearance or describe the behavior of any cocks identified as immature. The volume of evidence which I have collected indicates that Scott's heteroclite cocks were actually unrecognized immature males which had not yet attained full sexual maturity. Cocks which exhibited similar behavior on Eden Valley strutting grounds were considered to be normal in all respects.

Scott also has recognized the presence of heteroclite hens on the strutting grounds, and described their anomalous malelike behavior. Similar observations were made during the course of this study. Two hens assumed a fighting position similar to that of domestic roosters and struck at each other. On other occasions hens were observed in prolonged attempts at mating. During one of these mating attempts on the part of two hens, a cock mounted the top bird, and then another cock proceeded to hop onto this pile of three birds. A male was seen to break up a mating attempt by two hens. Rarely a hen would perform the strutting act. This type of hen behavior was not studied other than described. It was considered to be an unusual pattern of behavior for hens, but in the physiological sense was perhaps not too abnormal.

Scott's study of the mating behavior of the sage grouse was carried through an entire mating cycle. He worked out the social behavior of groups of birds composing a social organization of approximately 800 individuals on one large strutting ground. Scott concluded that an extraordinary system of polygamy prevailed in which dominance in males was based on fighting, bluffing, and strutting displays. Practically all mating took place on five mating spots, each not much larger than an ordinary room. Each spot was occupied by a more or less compact group of hens, by a dominant master cock that did most of the mating, by one major rival or "subcock" that took over some matings under certain conditions, and by several guard cocks that surrounded the hens and aided in keeping away intruders. The guard cocks rarely were allowed to mate. The remaining cocks were on widely distributed locations, singly or in pairs. Of 174 observed matings, dominance of cocks was observed in 154. Of this number, 114 (74%) were by master cocks and 20 (13%) by subcocks (5 by guard cocks, and 15 by outside cocks) under conditions where the system tended to break down. Mating took place only upon invitation of the hen.

Similar studies on the mating behavior of the sage grouse were made by this worker during the mating cycle of 1950 in Eden Valley. Intensive observations were conducted on a strutting ground which contained approximately one hundred males and an undetermined number of females. During the period from March 14 through June 5, a total of thirty-five mornings was spent in observation of courtship activities on this one strutting ground. Most observations were made from a photographic blind erected near the center of the strutting area; on a few occasions during the latter part of the season a pickup truck was utilized for this purpose. Adult cocks established both individual and group territories at an early date. By marking cocks and determining patterns of density and spacing, it was established that males occupy the same general positions on their strutting grounds on successive mornings. The fighting, bluffing, and strutting displays described by Scott are performed by

Cocks commence strutting activities in January on winter roosting sites

A few cocks move onto snow-covered strutting grounds in late February

The immature cock may be recognized by the narrow white band on the cape

The size of strutting ground plots varies from a few hundred square feet to several acres, depending upon the number of males utilizing the grounds

During a phase of their territorial defense a pair of cocks may engage in extended clucking duels prior to actual combat

Cocks defend individual strutting territories by bluffing displays and actual combat

The dominant cock is frequently surrounded by several "guard cocks"

The majority of matings take place during morning twilight

Hens appear on strutting grounds in large numbers in early April and mill around on the mating spots

In mid-April hens concentrate on strutting ground mating spots amidst dominant cocks

A dominant cock actually performs the majority of mating acts on a given mating spot

Hens signify their readiness for mating by squatting in front of a cock

A satiated cock may stand on a hen's back for several seconds after the mating act

Peak of mating activity occurs during the third week of April in Eden Valley

Swishing sounds are produced by wings rubbing against the stiff, white feathers on the cape

Grouse in Eden Valley terminate seasonal strutting activities in late May

Sage grouse cocks on winter range near Eden Valley

Spring snowstorms seldom disrupt strutting activities

Cocks utilize the same strutting ground throughout the breeding season

An immature male cautiously approaches the strutting domain of an old cock

A 7 lb. adult cock trapped on strutting ground—May 16, 1951

Yearling cock (left) and adult cock trapped on strutting ground
—May 10, 1950

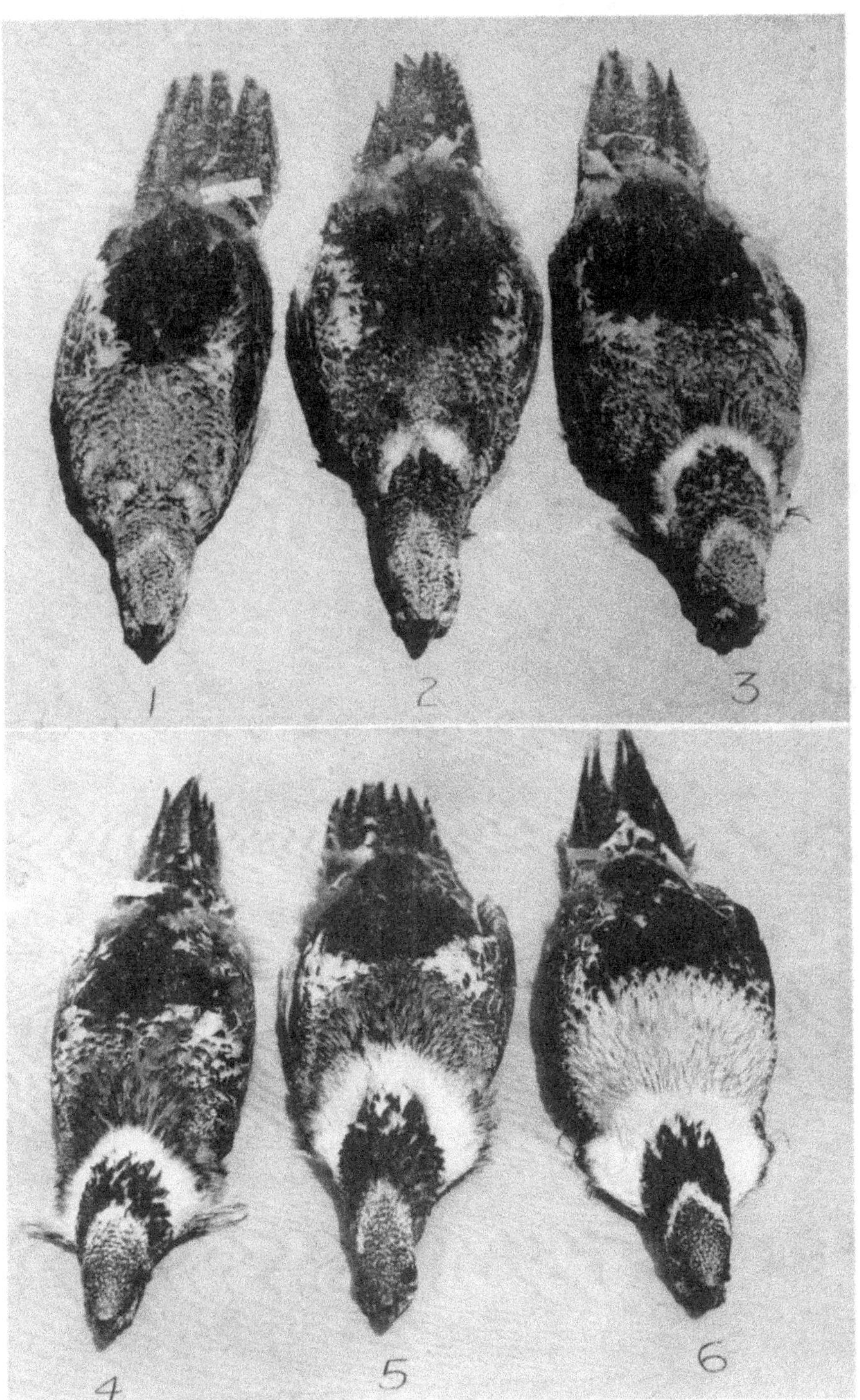

Development of the breeding plumage in the immature cock:
1—December; 2—January; 3—February; 4—March; 5—April;
6—April (old cock)

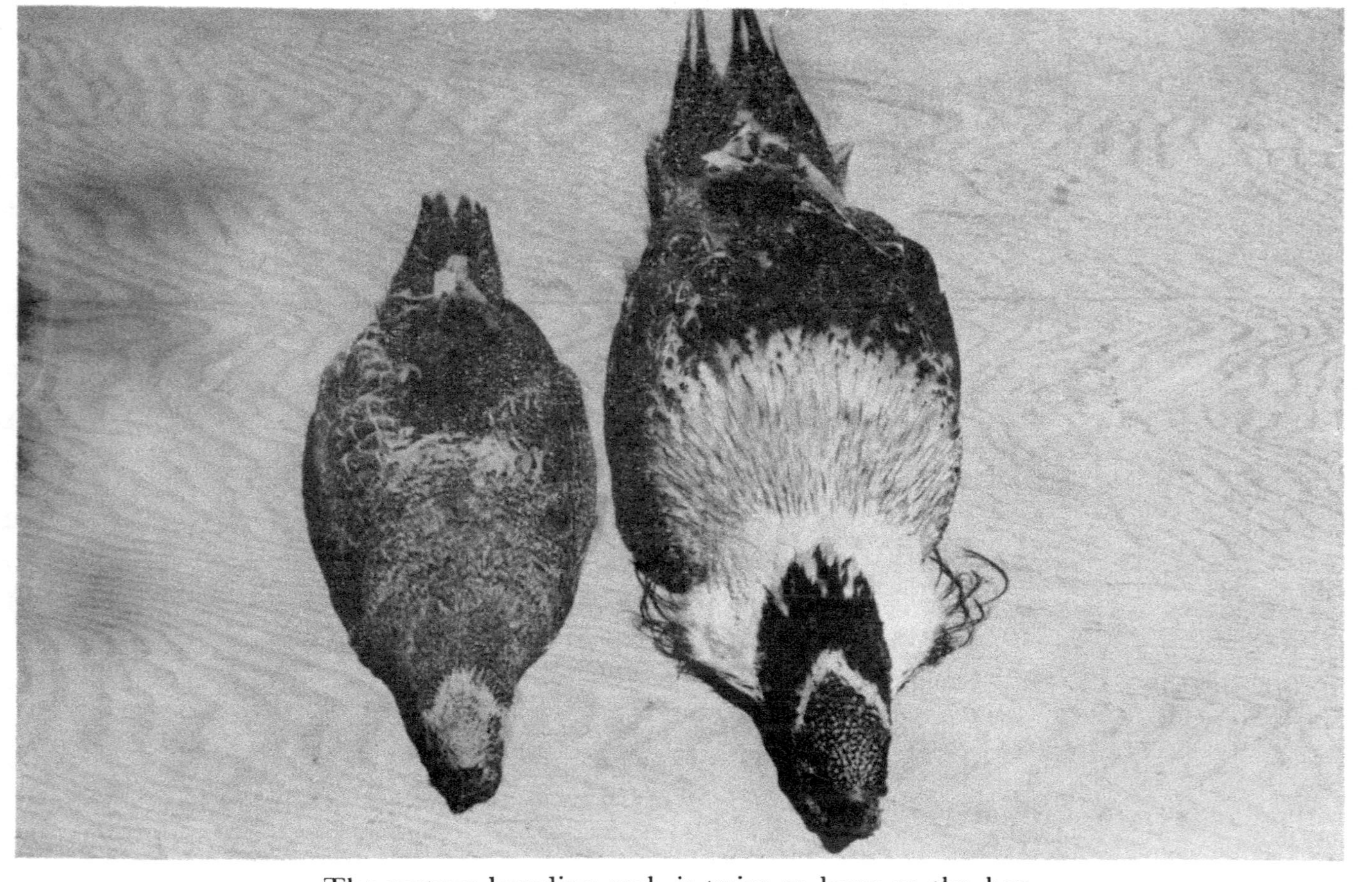

The mature breeding cock is twice as large as the hen

all of the adult males, but are significant in dominance behavior only within a small group of males, or commonly between pairs of birds. There was no pronounced movement of subdued birds, or renewed conflict for territory involving a new pair of cocks. The same cocks within a certain group territory engaged their associates in fighting and bluffing bouts at regular intervals each morning of the strutting season. There was no withdrawal to new areas on the part of vanquished birds, but merely a retirement of combatants to their respective strutting spots to await renewed combat at a future time. These fighting displays were vigorously waged, and many cocks were observed with bloody wings and blood-spattered breasts. Immature males which daily wandered onto the strutting grounds in small numbers were quickly contested. They put up no defense, and were easily driven off by adult cocks.

It was customary for cocks to move onto their strutting ground prior to the onset of morning twilight. Commencing with the appearance of large numbers of females at the morning display, generally about April 1, the males began to occupy the grounds during the evening. This evening occupancy began shortly after sunset in Eden Valley; it was of less magnitude and intensity than the morning display, and involved fewer birds. Following the evening display, the birds retired to nearby roosting areas. On bright moonlight nights during the peak of strutting activities it was common for some males to remain on their strutting grounds throughout the night, engaging in sporadic nuptial performances.

Cocks invariably flushed from the strutting ground as the observer entered the blind a few minutes before morning twilight. However,, within twenty to thirty minutes the entire assemblage would return and resume their strutting activities as if they had not been disturbed. If the birds were flushed during morning twilight or following sunrise they did not return until the next display period. Strutting grounds could be approached and effectively observed at close range in a motor vehicle at any time except during the waning stages of daily strutting activities. The birds became listless and flushed readily an hour or more after sunrise. All counts were made from a truck during the period from one-half hour before sunrise until one-half hour immediately after sunrise. Strutting ground observations in 1949 had indicated that the maximum numbers of males are present during this period and that counts obtained at any other time of the day are poorly representative of the total numbers of males utilizing the grounds.

Under conditions of climate and elevation similar to those existing in the upper Green River Basin, strutting ground censuses should be confined to the period from mid-April to mid-May. A series of counts on males taken at this time will be closely representative of maximum numbers utilizing the · grounds during the breeding season. In other areas of sage grouse range, census periods should be modified according to variations in elevation and climate.

A few hens first appeared on the Eden Valley grounds in mid-March, but were inclined to remain on the periphery at this early date. By the first week of April, females were visiting the grounds in numbers approaching that of males. During the forenoon display the hens flew onto the strutting grounds coincident with the appearance of morning twilight. Females habitually selected areas in the center of a strutting ground on which they would concentrate in groups of a few to several dozen birds, awaiting the actual mating period. Several of these prospective "mating spots" would be located on a large strutting ground, and the same spots would be utilized by groups of hens during successive seasons. During the morning twilight period hens alternately rested and milled around within the confines of these mating spots, but by sunrise or shortly thereafter they began to disperse into the surrounding brush.

From a position in a photographic blind placed in the midst of strutting activities, I had one of these grounds under daily surveillance from April 4 to April 22, 1950. Although females were present in large numbers, the first actual mating did not occur until April 14. During the days that followed, matings were observed with increasing regularity until April 20, after which time breeding activity declined rapidly.

Ten hens appeared at a mating spot on April 8 and milled around among a group of displaying cocks. Fourteen hens were present on April 9. On April 10 there were sixty-four hens grouped on two separate mating spots. One of these groups dispersed fifteen minutes prior to sunrise. The other group remained intact until thirty minutes after sunrise at which time they moved into a different concentration of strutting males. Only thirty-two hens showed up on previously identified mating spots on April 13 and twenty-four hens appeared on April 14. None of the hens indicated a willingness for mating nor were any matings observed prior to April 14. The variation in the number of hens which appeared on the grounds may have been the result of an extensive migration through the Eden Valley area which was apparent from other field observations conducted at this time. It has not been determined whether hens visit several strutting grounds before selecting a certain one for the mating act.

The mating attempts noted on April 14 and 15 were not completed because the dominant cocks were knocked from the backs of the hens by other cocks. Several other hens indicated their readiness for mating by squatting in front of displaying cocks and by spreading their wings against the ground. Eleven separate and complete matings occurred on April 16 of which eight were accomplished by a single cock (color marked). This bird was attacked by other cocks and knocked from the mounting position on hens on six separate occasions. Thirty hens were present at this mating spot on April 16.

On April 18 a total of eighteen mating attempts was consummated. Cocks were disrupted on seven of these attempts by the interference of other cocks. The bird which accomplished eight of the eleven matings on April 16 also performed seven of the eighteen mating attempts observed on April 18; three other cocks performed the balance of the matings. Approximately thirty hens were present on this mating spot on April 16, 17, and 18. Two banded females were present on April 15 and 16, although it was not possible to determine whether the same birds appeared on consecutive days. Still another hen, banded during the 1949 nesting season and whose complete history is listed in the nesting discussion, arrived on April 16, and was not seen again on the strutting ground. The number of cocks in the vicinity of the mating spot noticeably increased during the actual period in which hens were congregated there for mating. Most of these cocks were very active in their courtship displays but few were seen to take part in the mating performances. In most cases the groups of hens dispersed from mating spots before sunrise. Several of these scattered hens were thereupon mated by cocks stationed apart from the central mating spots. Breeding activity continued at high intensity on April 19 and 20. Matings were again accomplished by four separate cocks on the mating spot previously studied. One of these cocks made thirty-five separate mating attempts with the same hen. On twenty of these occasions he was literally knocked from the mounting position by other cocks. He soon became incapacitated and toward the end of this marathon he simply stood or sat on the back of the hen, making no attempt at mating. This cock was blood-splattered on the wings and breast due to repeated blows from other cocks. A similar procedure was again witnessed on April 20. On this date a cock, presumed to be the one observed the previous morning, began mating with a hen at 5:00 a.m., and at 6:15 a.m. was still engaged in mounting the same female.

The intensity of mating had abated considerably by April 21, although a few females were seen on the strutting grounds throughout the remainder of the strutting period. On April 22 a cock mated with three hens in succession, stopped to fight with another cock, then proceeded to mate with still a fourth hen.

There was little doubt that sage grouse males exhibit patterns of social dominance on their strutting grounds. At the height of the breeding period most females select favored spots for mating on their initial appearances upon the strutting grounds. Hens evidently recognize the dominance of certain adult males, and respond accordingly in their selection of mating spots. Within these mating spots a few males assume a dominant role in courtship and actually perform the majority of matings. The females quickly disperse from these mating spots as full daylight arrives, or as the dominant cocks become preoccupied with defending their harems against intruders. Matings occasionally took place at other localities on the strutting ground. No actual matings were observed during the evening displays, nor during periods of darkness. On the evening of May 28, 1951, while I was concealed in a blind on a strutting ground in Jackson Hole, Wyoming, a hen repeatedly indicated her willingness to mate by squatting in front of a cock. As the cock approached to mount, the hen invariably moved slightly, with the result that the actual mating act was never performed. It seemed likely that mating would occur during the evening display, although hens seldom appeared on the grounds at that time.

Strutting grounds varied tremendously in size, and at the peak of the season supported from a few to several hundred birds. Various age classes undoubtedly were present on every strutting ground, though immature males were the only birds readily distinguishable. Older male age groups possibly might be identified by their size and the character of their nuptial plumage. Hens appeared indistinguishable in the field, irrespective of age groups involved.

Intensive studies on sage grouse behavior and dominance patterns were confined to a small area and to a single mating spot on a strutting ground. It was exceedingly difficult, at the best, to observe and follow the movements of individual birds, identify and keep them segregated, and interpret the behavior and dominance patterns exhibited on even a small area of a strutting ground. The birds, furthermore, had to be marked either artificially or naturally in order for a person to recognize individual grouse and to record their actions and movements accurately.

In the complete absence of marked birds and a lack of knowledge of age groups, it is my considered opinion that any attempt to describe the individual behavior, movements, and dominance patterns displayed by large numbers of sage grouse on their strutting grounds would be extremely hazardous.

A wide variety of topographical features and land-use areas are selected by sage grouse, and utilized as locations for strutting grounds (Table 19).

Heavy use of strutting grounds by sage grouse over a long period of years served to create permanent openings in the vegetation as a result of their extensive trampling and feeding activities. Often they take over abandoned man-made clearings, sites of old saltlicks, sheep corrals and loading chutes, bedgrounds,

TABLE 19

Types of Land Areas Utilized as Strutting Grounds by Sage Grouse

Type of Land Area	Number of Grounds
Windswept ridge and exposed knolls	11
Flat sagebrush areas with no openings	10
Bare openings on relatively level land	7
Alkaline flats	5
Small water-collecting basins	4
Adjacent to streams	3
Scattered dirt mounds from abandoned rodent or ant colonies	2
Dry lake bottoms	1
Abandoned homestead sites	1
Excavations around gravel pits	1

dips, sheep wagon camps, etc. They also frequently utilize airstrips as sites for strutting. For years, sage grouse have strutted on the Western Airline's runway at Jackson, Wyoming, as well as on Frontier Airway's field at Riverton, Wyoming. Deputy Game Warden Norbert Faass reported that forty sage grouse cocks were observed strutting on the Fulkerson Airport about 5 miles northeast of Gillette, Wyoming, during April, 1947.

The size of strutting grounds varies from plots containing from a few hundred square feet to plots covering several acres, depending upon the number of males utilizing the grounds. When denuded or exposed areas with stunted vegetation are used, it is customary for only 30 or 40 per cent of the cocks to strut on the opening, with the remainder of the birds performing their display in the vegetation around the edge of the opening.

Four pairs of strutting grounds on the study tracts were spaced within 0.25 mile of each other; two pairs were 0.5 mile apart; and five pairs were about 1 mile apart. The remaining grounds were all spaced more than a mile from the next adjacent strutting area.

Many smaller grounds in close proximity to larger strutting areas apparently originate as offshoots from a larger parent ground. On these occasions, immature males are probably driven from old established grounds by the dominant males, and subsequently move into new areas where they are unmolested to establish strutting grounds of their own. An example of this behavior occurred in Jackson Hole during the breeding season of 1951. A new strutting ground was reported by local residents for the first time along the highway southeast of the Western Airline's Airport. Upon investigation, it was noted that this newly established courtship area was located about 0.25 mile from a permanent strutting area. At least ten of the eleven cocks on the new strutting ground were immature birds.

Due to the abundance of water over the study areas in the spring, it was difficult to determine whether availability of water was associated with the selection of permanent strutting sites. Some of the strutting grounds were located as far as 2 miles from open water, whereas others were within a few yards of streams or ponds. Sage grouse are powerful flyers with good endurance. Daily movements of several miles to water are not unusual.

In areas of high grouse concentrations, strutting grounds generally are located at random. This is in contrast to the location of nesting areas which are concentrated primarily in specialized regions on both study tracts. In an extremely large area, such as the Green River Basin, the amount and distribution of available water seems to be an important factor contributing to high nesting populations of sage grouse in localized areas.

Within the Eden Valley district strutting grounds are ordinarily located 1 to 3 miles from habitations. Three grounds are less than a mile from human dwellings, and one is located about 300 yards from an abandoned CCC camp. Strutting grounds are commonly traversed by graded roads and sheep wagon trails. Portions of two grounds frequented by from 75 to 100 cocks are actually within the right of ways of state and federal highways.

A strutting ground site, unique in its location, is situated on the dry, sandy bottom of the Eden Valley reservoir. As the west cove of the reservoir fills with water in the spring the birds gradually retreat toward the high waterline, finally being displaced completely from all strutting territory utilized at the beginning of the season. At no times were strutting activities disrupted, and when the reservoir was filled, the cocks were still strutting along the shoreline.

Under most circumstances, the birds seem adaptable to physical changes produced on their strutting grounds. Contrary to popular thought, the alteration or obliteration of a strutting site does not result in a degeneration of that particular segment of the sage grouse population as far as future reproductive capacities are concerned.

James R. Simon related in a letter that on May 15, 1950, at 6:30 a.m. he observed fifty-one sage grouse cocks strutting in a wild hay meadow one mile west of Warren Bridge, Sublette County, Wyoming. They were decidedly more active than birds he had observed a few days earlier on strutting grounds at Farson. Impressive to him as the ultimate in variety of strutting areas was the fact that half the birds were working in water about an inch deep over the flooded meadow. The other cocks were strutting on dry ground.

Description of the Strutting Display

Several persons have described in their distinctive manner the various phases of the sage grouse cock's strutting display. It is generally agreed that in the preliminary phase of strutting the male spreads the tail feathers in a semicircular pattern, erects the filiplumes over the head to form a lyre-shaped figure, raises the head, partially inflates the air sacs, and thrusts the wings slightly forward with the tips of the primaries almost touching the ground. The strutting performance continues as the male inhales air with a gulping motion to further inflate the pendulous air sacs of the esophagus; he then quickly takes several steps forward, stops abruptly, and with a slight rotary motion the wings are elevated and retracted, causing the leading edges of the primaries to scrape against the scalelike feathers of the cape to produce an audible swishing sound. The latter motion is repeated three times in rapid succession. On the final retraction of the wings across the cape, the entire air sac region is fully inflated and bouncing on the breast of the bird. During the peak of strutting activities the entire performance just described may be repeated as often as twelve to fifteen times per minute.

According to Honess and Allred (op. cit.) the resonant plopping sound is produced by the sudden contraction of the neck muscles which traps air in the distended portion of the esophagus, causing the membranes of the bare areas to vibrate. The bare areas act as a sonorous membrane comparable to the skin on a drum.

The plopping sound resembles that which emanates when a rock is dropped into an old well, not being particularly loud, but possessing high-carrying qualities. It is not produced as a result of collapse of the air sacs as stated by several observers.

Strutting Ground Observations in Jackson Hole

A small isolated sage grouse population has survived since primitive times on the sagebrush-covered outwash terraces bordering the Snake River in Jackson Hole, Wyoming. Homestead claims, community settlements, and reclamation activities have eliminated a large portion of the original sagebrush-grass habitat available to wildlife in this intermontane valley. Most of the remaining sagebrush areas which furnish summer range for sage grouse are included within the boundaries of the recently enlarged Grand Teton National Park.

A rather intensive coverage of this entire area was made in 1947 and 1948 in an attempt to locate all sage grouse strutting grounds in the Jackson Hole region. Eight separate grounds were initially located (Figure 13). Since 1948, strutting ground censuses have been made annually in this area, concentrating on six of the largest grounds. These counts have revealed only the slightest yearly fluctuations in the Jackson Hole sage grouse population since 1948 (Table 20). Due to time limitations, the yearly strutting ground counts were usually based upon a single morning's observation, and hence may not necessarily represent maximum numbers of cocks present on the strutting grounds during that year. Sage grouse are fully protected in the Jackson Hole region, although some illegal hunting has been reported on areas formerly outside the jurisdiction of the National Park Service.

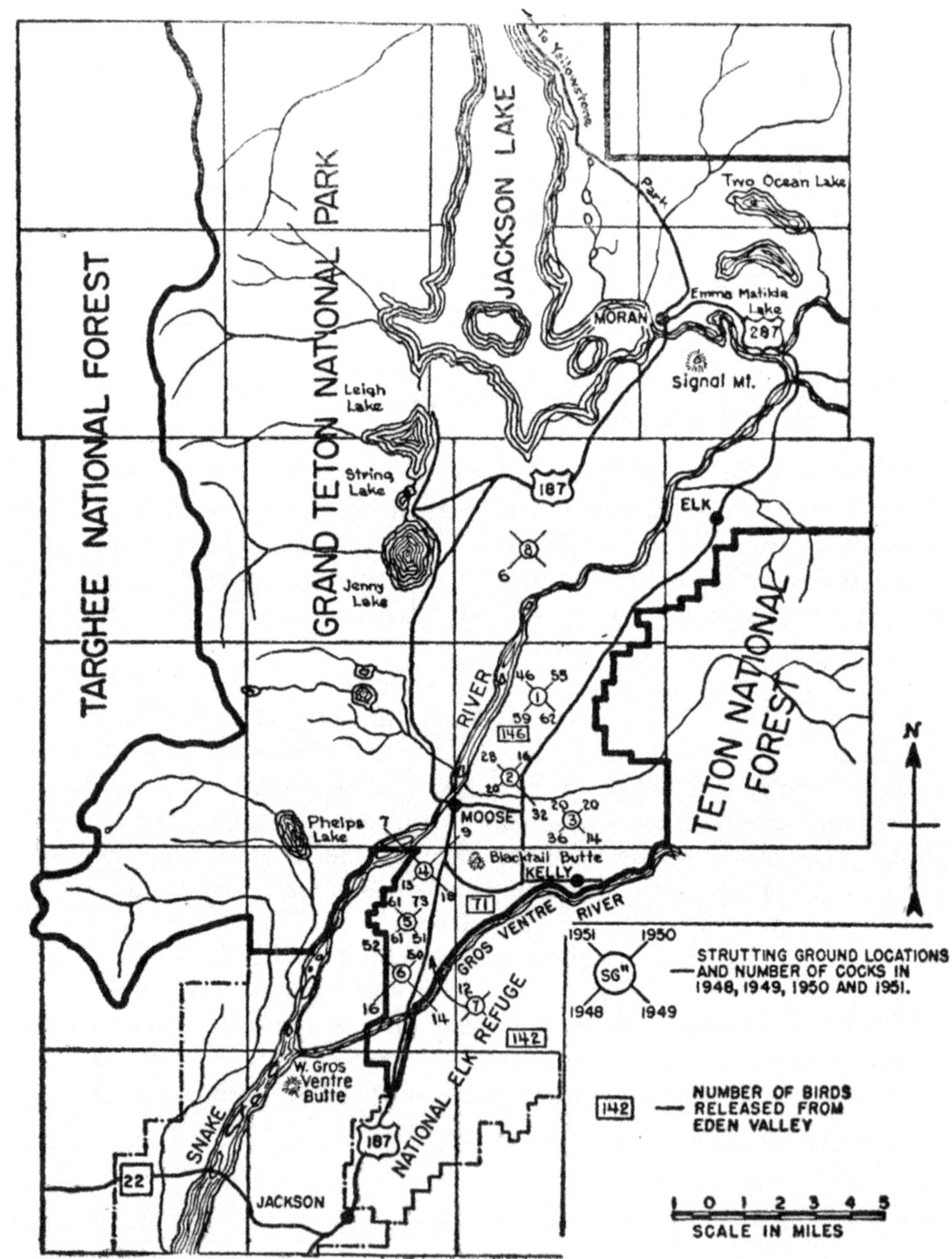

FIGURE 13.   LOCATIONS OF SAGE GROUSE STRUTTING
GROUNDS AND RELEASE SITES IN JACKSON HOLE, WYO.

TABLE 20
Censuses of Sage Grouse Cocks on Strutting Grounds in the Jackson Hole Region

Location of Strutting Ground	1948	1949	1950	1951
1. Antelope Flat	59	62	55	46
2. Old Blair Homestead	20	32	16	28
3. Warm Springs near Kelly	36	14	20	20
4. 3 Bar H Road	13	18	9	7
5. Western Airlines Strip	61	51	73	61
6. One Mile S. E. of Airstrip	16	14	50	52
	205	191	223	214

Structural and Adaptive Features of the Sage Grouse

Form of the Body

In comparison with other upland game birds, sage grouse are heavy-breasted, long-bodied, short-legged, and strong-winged birds.

Seasonal Weights

In contrast to the seasonal weight trends exhibited by certain other game birds, sage grouse of both sexes attain their maximum weights, not in late fall, but in the early spring, following the winter months (Table 21). The average weights of thirty-three mature males and sixteen mature females examined during the spring breeding season was 5 lb. 15 oz. and 3 lb. 1 oz., respectively.

The heaviest weight recorded was 7 lb. on an adult male trapped on a strutting ground near Pacific Springs, May 18, 1951. Weights of mature males during the strutting season varied from 5 lb. 2 oz. to 7 lb. Weights seemed to be related to the ages of the birds and to their state of sexual development. The maximum weight recorded on an immature male was 5 lb., taken on April 19, 1950. By midsummer, these immature birds (now yearlings) had dropped in weight to an average of 4 lb. 4 oz. as compared to an average weight of about 4 lb. 9 oz. for fully-matured cocks at the same period.

The maximum recorded weight for a female was 3 lb. 6 oz. taken from a bird killed on the highway, April 7, 1951. Weights of females during the breeding season average slightly over 3 lb., decreasing to 2 lb. 9 oz. during the postnesting period. Four immature females collected during the spring of 1951 averaged 2 lb. 15 oz. Three yearling hens collected on June 14, 1951, averaged 2 lb. 11 oz., whereas the average weight of eight adult hens taken at the same time was 2 lb. 14 oz.

TABLE 21

Average Monthly Weights of 322 Adult and 263 Young Sage Grouse from the Upper Green River Basin—1949 to 1951

Month	No. Specimens		Aver. Wgt. in Lbs.				Av. Wt. in Grams	
	Male	Female	Male		Female		Male	Female
Adult								
			Lb.	Oz.	Lb.	Oz.	Grams	Grams
March	2	2	6	4.0	3	0.5	2,835	1,374
April	2	1	5	14.5	3	6.0	2,679	1,531
May	29	----	5	15.5		----------	2,707	-------
June	4	11	4	14.3	2	13.2	2,200	1,181
July	97	31	4	6.9	2	8.3	2,010	1,142
August	31	26	4	9.1	2	9.8	2,072	1,185
September	----	----	----------		----------		-------	-------
October	25	57	4	8.5	2	9.3	2,044	1,171
November	----	----	----------		----------		-------	-------
December	1	2	5	9.0	3	1.0	2,523	1,389
January	----	----	----------		----------		-------	-------
February	----	1	----------		2	14.0	-------	1,304
Young								
			Lb.	Oz.	Lb.	Oz.	Grams	Grams
June	2	3	0	7.7	0	4.0	218	113
July	66	52	1	8.3	1	2.7	689	530
August	29	30	2	4.8	1	10.2	1,043	742
September	----	----	----------		----------		-------	-------
October	54	15	3	6.6	2	3.7	1,548	1,012
November	----	----	----------		----------		-------	-------
December	2	----	4	7.0		----------	2,013	-------
January	1	----	4	7.0		----------	1,928	-------
February	1	----	4	10.0		----------	2,098	-------
March	1	1	4	8.0	2	15.0	2,041	1,332
April	2	----	4	13.5		----------	2,197	-------
May	3	----	4	15.0		----------	2,240	-------
June	2	3	5	0.0	2	11.0	2,268	1,219

TABLE 22

Weights of 33 Mature Sage Grouse Cocks Taken During the Breeding Season at Eden Valley—1949 and 1950

Date	Weight	Date	Weight	Date	Weight
Mar. 5	6 lb. 6 oz.	May 9	6 lb. 2 oz.	May 10	5 lb. 9 oz.
Mar. 19	6 3	May 9	6 2	May 11	5 12
April 20	6 0	May 9	6 1	May 11	6 3
April 25	5 13	May 9	6 7	May 11	5 13
May 1	6 2	May 9	6 9	May 11	5 11
May 5	5 2	May 9	5 14	May 13	6 12
May 5	5 9	May 9	6 1	May 14	6 3
May 5	5 13	May 9	5 12	May 14	5 8
May 8	5 11	May 9	6 7	May 14	6 9
May 8	5 12	May 9	6 1	May 15	5 5
May 9	5 lb.10 oz.	May 10	5 lb. 8 oz.	May 18	7 lb. 0 oz.

Following the completion of the postnuptial moult and a slackening of the heavy physiological demands of courtship, nesting, and moulting, the weights of both sexes increase a few ounces throughout the fall period. By midwinter the birds are again approaching their highest weight levels of the year.

Live-trapping operations on the strutting grounds afforded an opportunity to obtain weights on about three dozen cocks in the breeding condition (Table 22). Later in the season, several of these cocks were subsequently recaptured, and weighed during live-trapping activities in alfalfa fields in Eden Valley. The weight losses of these banded birds, three to four months after breeding season, varied from 1 lb. 13 oz. to 1 lb. 1 oz. (Table 23).

Girard (op. cit.) weighed a separate cock and hen taken in Sublette County, Wyoming, during each week of July and August, 1934. The average weight for eight cocks was 5 lb. 10.4 oz., and for eight hens was 3 lb. 2.4 oz. These average weights are substantially greater than maximum weights obtained during the present study on a series of adult birds in Eden Valley. The maximum weight recorded during the same months for adult males in Eden Valley was 5 lb. 6 oz., with an average weight of 4 lb. 9 oz. The maximum weight for adult females in Eden Valley during July and August was 2 lb. 14 oz., with an average

TABLE 23

*Weight Changes of Individual Male Sage Grouse
After the Breeding Season*

Mature Cock No. 2589—May 11, 1950—6 lb. 3 oz.
                    Aug. 8, 1950—4 lb. 11 oz.
                                1 lb. 8 oz. loss in weight

Mature Cock No. 2595—May 13, 1950—6 lb. 12 oz.
                    Aug. 8, 1950—4 lb. 15 oz.
                                  1 lb. 13 oz. loss in weight

Mature Cock No. 2591—May 11, 1950—5 lb. 13 oz.
                    Sept. 2, 1950—4 lb. 12 oz.
                                  1 lb. 1 oz. loss in weight

Yearling Cock No. 3393—July 21, 1950—4 lb. 1 oz.
                    Aug. 9, 1950—4 lb. 0 oz.
                                          1 oz. loss in weight

Juvenile Cock No. 3317—July 17, 1950—1 lb. 7 oz.
                    Oct. 4, 1950—3 lb. 7 oz.
                                  2 lb. 0 oz. gain in weight

weight of 2 lb. 9 oz. The reasons for the differences in weight recorded in this study and Girard's are unknown. Except for a slight increase in elevation and proximity to the mountains, the Sublette area is comparable to the Eden Valley tract. If these weights are truly representative of two segments of Green River Basin sage grouse populations, birds in the foothill region may remain in better condition during the summer months than birds in habitat on the drier central portions of the Basin. However, there is considerable doubt that this is true, since two fully matured cocks collected at South Pass on June 29, 1951, at an elevation of almost 8,000 feet, weighed 4 lb. 14 oz. and 4 lb. 13 oz.

For most animals the winter season is the most critical period of the year and the most taxing upon their physiological systems. Sage grouse which I have observed in the Green River Basin came out of the winter months in their finest physical condition of the entire year—fully conditioned for an arduous three-month courtship and reproductive season. It would be interesting to know whether other animals utilizing sagebrsuh on their winter range exhibit similar weight relationships. Antelope, mule deer, domestic sheep, and many smaller mammals consume large amounts of sagebrush in their winter diets. For purposes of comparison, seasonal weights were unavailable on the other species of prairie grouse which have courtship performances similar to sage grouse. An examination of the seasonal weight trends of prairie chickens and sharp-tailed grouse might reveal the significance of the weight relationships displayed by sage grouse. The nature of the climate and the pattern of plant growth in semiarid regions may also be responsible for alterations in the biological relationships of desert animals from that existing in more humid environments. The high affinity of sage grouse and antelope for cultivated alfalfa, clover, and other ranch crops in the dry, late summer periods may be further evidence of the varying effects of physiography and climate upon the seasonal aspects of animal physiology.

Measurements

Due to the large variation in size between the adult sexes, measurements of body parts may be used effectively at an early age to determine the sex of sage grouse. In the complete absence of plumage characters, carcasses of mature birds may be readily identified as to sex by checking measurements of skeletal characters such as the head, tarsus, toes, and wing chord. Measurements of adult and immature birds were taken in October in order to indicate the approximate range of sizes between sage grouse of both sexes at this season (Table 14). After an examination of several hundred juvenile birds, the actual size differences could be converted into gross appearance differences. Sexing of young birds soon developed into a rapid technique, which could be used reliably in large-scale handling operations, e.g., game checking stations and live-trapping and banding.

Flight

The semidesert habitat has produced a bird which is highly mobile and capable of long distance flights at comparatively high altitudes. Sage grouse possess musculature features of typically migratory birds such as ducks, geese, and hawks. The breast, or flight muscles, exhibits the deep, blood-red color characteristic of muscle tissue highly resistant to fatigue. Most of the other upland game birds possess the so-called white breast meat. Muscles of this latter group of birds are adapted for powerful short distance flights, but, on the other hand, are quickly fatigued.

Nostrils

The external nasal openings (nostrils) of all grouse species are completely feathered. As an aid in surviving severe winter seasons with accompanying snow storms, ground blizzards, and sub-zero temperatures, the inner nasal openings of sage grouse are provided with a type of valve structure featuring a cartilaginous receptacle and a plug. When this structure is closed, the nostril is effectively sealed against cold winds and blowing snow. During the most severe winter storms sage grouse may be observed feeding unconcernedly on wind-swept ridges and buttes.

Neck Region

Accompanying the change in plumage and gonad size at the onset of the breeding season is an enlargement of the musculature, connective tissue, and esophagus in the neck region of sage grouse, particularly in males.

Honess and Allred (op. cit.) have described the structure and functions of the neck muscles in the strutting performance of sage grouse cocks. Additional studies on the anatomical and physiological changes of the air sac region of sage grouse have been made by Clarke, Rahn, and Martin (op. cit.). Associated with the increase in size of the testes they found a significant increase in esophagus volume and in the size of the bare patch areas. Slight increases in esophagus volume as well as in bare areas were also noted in adult hens during the breeding season. Esophagus capacity of the adult male in the breeding condition was about twenty-five times that encountered in the adult non-breeding male.

Digestive System

The sage grouse possesses the typical fowl-like digestive system, although there are many persons who will proclaim vociferously that the bird does not have a gizzard. This belief arises from the fact that the sage grouse stomach is a thin-walled muscular organ and membranous in structure, as compared with the heavy-walled muscular organ found in most other upland game birds and domestic fowl. Dissection of a sage grouse stomach revealed no significant differences in structure from that encountered in other gallinaceous forms acknowledged to possess a gizzard. The gizzard of a sage grouse exhibits less musculature than is found in the typical grain-eating game bird or domestic fowl; otherwise the structure is homologous, even to the presence of a horny, inner lining. A gizzard is defined by Webster as the second stomach of birds, having thick walls and a heavy lining for grinding food. The *American Illustrated Medical Dictionary* defines a gizzard as the strong muscular stomach of a bird.

The diet of sage grouse consists primarily of herbaceous leafy material supplemented by insects. Cereal crops and other hard seeds are normally not a feature of the semiarid sagebrush plains and consequently have not been a factor in the development of the sage grouse digestive system. Neither is the intake of gritty substances a feature of sage grouse feeding activities in the sense that it is used as an aid in the grinding processes of digestion.

Toes

Early in the fall, a comblike fringe of minute appendages appears along each side of all three toes of the sage grouse. These are shed late in the following spring. Each tooth of this fringe is provided with a hollow core which fits over a knob of epidermal tissue on the toe proper. Each tooth drops off in the spring in a manner not unlike the shedding of the horns of the pronghorn antelope. These fringes on the toes are distinct aids in walking over the snow and are popularly called "snowshoes." Other grouse of snow regions possess similar adaptations for winter survival. The inner side of the nail on the middle toe of each foot has a somewhat enlarged, sharpened, and nonserrated edge which apparently is used by the bird when scratching various parts of the body.

Plumage

In addition to the remarkable pattern of protective coloration, making the sage grouse highly adaptable for survival in a sagebrush habitat, the plumage exhibits other features unique in their function. The leading edges of the primaries are finely serrated and assist in producing the sound from wings swishing against the cape as the cocks go through their strutting performance. The tarsus of the sage grouse is completely feathered to the base of the toes, except for a narrow strip on the ventral surface.

Albinistic sage grouse occur very rarely. A local taxidermist has reported mounting one such specimen. During the early years of the sage grouse investigations in the Pacific Creek area, field crews spent considerable time looking for a cream-colored sage hen reported by a sheep herder, but had no success.

Behavior Patterns of the Sage Grouse

The reactions of sage grouse to various stimulating factors in their environment are important not only from the standpoint of a knowledge of their collective behavior patterns; they have practical application in a consideration of their prospects for survival in a habitat constantly being altered by man's land-use activities. They must also be analyzed and reviewed in establishing management plans for the bird. The success of census techniques, live-trapping and transplanting programs, protective measures, hunting seasons, and habitat improvement programs are vitally dependent upon proper knowledge and evaluation of sage grouse behavior patterns.

Disposition and Temperament

Sage grouse possess few qualities of individuality. They are strongly inclined to be gregarious at all seasons of the year, even to the extent of nesting in close proximity to one another. Communal courtship and nesting activities emphasize the colonial nature of the species. Responses to external stimuli are exhibited by mass reaction rather than by initiative on the part of each bird. The success experienced in large-scale live-trapping operations is dependent upon the masses following the movements of a few birds. The actions of an individual bird on the strutting ground will motivate the departure of the entire group en masse.

The average hunter deplores the apparent lack of intelligence exhibited by sage grouse, and is likely to assign the species to the "fool hen" category. During the spring and summer months the birds are quite tame and are reluctant to avoid contact with man, either in the ranch yard, on the highway, or before the poacher's gun. The heat of the desert summer and the accompanying moulting period contribute to the bird's sedentary nature and lack of desire to fly unless hard pressed. It is a species which has had little contact with civilization. This grouse thrives in essentially a wilderness area, and throughout the cold fall and winter months is rarely seen by the "hot stove leaguers." Even during legal shooting seasons large segments of the sage grouse population in inaccessible regions sustain no hunting pressure whatsoever. There is little wonder that the species seldom displays much fear of man. In this respect it is similar to the spruce grouse, blue grouse, and ptarmigan. Actually, in Wyoming I have observed little difference between the sage grouse and ruffed grouse in their behavior toward man. The latter bird, under complete protection in Wyoming, is fully as tame as the most stupid "fool hen," and yet in many areas of the country the ruffed grouse is considered to be the king of upland game birds and the ultimate challenge to the sportsman's skill. Many of the potential sporting qualities of the sage grouse have been overlooked due to the long standing tradition of conducting hunting seasons in August and early September. Practically every state has recognized the imperative need for shooting all other upland game birds and waterfowl during the fall harvest period, yet few states have awakened sufficiently to the need for extending the same sporting chance to either the sage grouse or the sage grouse hunter. In the fall, the sage grouse is a distinctly different type of bird than it is in the summer. It becomes more wary, flies more readily and for longer distances, and in general exhibits more intelligence than is commonly displayed during the warm summer months.

Within their own flock organization sage grouse appear to be quite sociable. There seems to be no competition between individuals for the essentials of daily survival such as food, cover, or water. The behavior exhibited by both sexes on the strutting grounds indicates that dominance may be established in the flock organization. Endurance and strength for both walking and flying are well-developed. Birds in confinement or in a state of injury rapidly lose their vitality and death follows within a very few days. At times memory patterns seem to be highly developed and at other times their capacity for remembering seems to be relatively short-lived. Birds return to the same strutting areas year after year as well as to identical nesting sites (confirmed by observing banded birds). As soon as courtship activities are concluded in the spring, the cock birds in Eden Valley commence daily movements into the ranching areas for the purpose of feeding in the alfalfa fields. At the other extreme, several repeats were obtained on birds during strutting ground and alfalfa field trapping operations, indicating that they had either forgotten a previous trapping experience, or else were quite unimpressed.

Senses

Young sage grouse are precocial, and, therefore, have acquired a high degree of sensory development at birth. In response to the hen's actions and call notes they soon develop highly selective feeding habits and rapidly learn the art of concealment and escape from natural enemies. The entire courtship performance is based upon the ability of the sexes, particularly females, to hear and interpret the sounds of the strutting cocks. The system of reproduction has been developed to a high degree of efficiency on range featured by the vastness of the habitat and the wide distribution of the bird. The character of the male's nuptial plumage and spectacular courtship display further appeals to the various senses of the female. The sounds produced by strutting cocks being stimulated to renewed activity when recorded strutting sounds are played back to them from a sound-recording ting sounds are played back to them from a sound-recording machine set up on their strutting grounds.

The birds often assume an attitude of listening or looking,

atop a clump of sagebrush, elevated mound of dirt, or occasionally upon a hay stack. Hens are more frequently seen in these manners as they guard their broods, or reassemble scattered chicks. As flocks walk into the vicinity of traps placed along the edge of alfalfa fields, lone individuals act as sentinels, and seem to give clearance to the advancing birds which may have sensed danger and halted temporarily.

Adult birds possess well-developed senses of orientation and homing. Fully two-thirds of several hundred mature birds removed from Eden Valley in 1950 and released at distances of from 15 to 30 miles to the southward of the valley, returned to the general site of their capture within a few weeks after their release. Many of these birds were recovered on the same ranches from which they were trapped. Three hens released near Evanston, Wyoming, in the fall of 1949 were killed within a few hundred yards of the point of their capture in Eden Valley in 1950 and 1951, involving a return movement of more than 100 miles. Two other hens, released 30 miles south of Eden Valley in 1947, were recovered in the valley in 1950. A mature male caught on July 20, 1950, was released 15 miles to the southeast, and recaptured on August 8, 1950, about 1 mile from the original point of capture. An adult female performed a similar return movement within three weeks after capture. These records were obtained on birds originally trapped and banded in Eden Valley, and either were recaptured by trapping or were killed during the open hunting seasons in August of 1950 and 1951. Birds have returned to the locale of capture from points of release varying from a few miles to more than 100 miles distant. Releases have been made at distances up to 350 miles from Eden Valley during the past ten years. Other birds may have returned, but were not recovered since hunting seasons were not held until 1948.

Periods of Activity and Leisure

The daily routine of sage grouse follows closely the pattern of activity of most other diurnal animals. That is, feeding and drinking activities are confined mainly to the daylight hours of early morning and late afternoon. During the midday period birds frequently are more inclined to rest and to remain inactive.

The sage grouse feeding pattern has been closely observed in connection with live-trapping operations in alfalfa fields. This has been an important phase of trapping operations, since the birds are caught as they move into the fields for daily feeding activities. Normally the afternoon movement of birds into the alfalfa fields occurred subsequent to 4:00 p.m. Feeding activities ordinarily ceased by sunset or shortly thereafter. On exceptionally hot and dry days the birds may be delayed in their afternoon movement into the fields until 5:00 or 6:00 p.m. The forenoon feeding period commenced during early morning twilight and was of short duration. Usually the birds departed from the fields within one hour after sunrise. During the period between feedings they retired to areas from several hundred yards to 2 miles distant from the fields and spent the midday hours resting in the shade. Feeding activities in alfalfa fields were greatly restricted during periods of rainy weather. Birds avoided the rain-soaked alfalfa due to the drenching effect of water upon their plumage. In the vicinity of irrigation systems the period of daily drinking activities coincided with the feeding schedule.

With the onset of autumn killing frosts, Eden Valley birds again retreated to the vast sagebrush prairie where they resumed their normal feeding and drinking routines. The sagebrush-grass range types furnished an ever present supply of feed. Grouse obtained water by daily movements to streams, springs, and marshy meadows. There was a tendency for birds on the open prairie to select the taller sagebrush in draws, gullies, and along stream beds for the midday resting spots, in contrast to areas of smaller and more open sagebrush selected for night roosting locations. Selection of cover apparently involved a shade-relationship in the daytime. At night, provisions for detection of danger and escape from their natural enemies were paramount. The vegetation served no function other than that of protective cover. Birds neither roosted nor rested on the limbs of the sagebrush, but merely squatted within openings or in jack rabbit forms at the bases of individual bushes. They had no aversions to squatting on the ground between clumps of brush as evidenced by the accumulation of droppings found in these locations. Individual roosting spots were in close proximity to other birds. During the late fall and winter seasons, a single roosting area may encompass several dozen acres and involve many hundreds of birds. Local residents have reported to me that sage grouse on occasion dig into deep-crusted snow and roost in these open holes. Griner (op. cit.) reported the same behavior during his sage grouse studies at Strawberry Lake, Utah, and obtained photographs showing a series of the roosting forms in deep snow. He was of the opinion that the birds dug down to reach the tips of sage-brush leaves, although they had roosted in some of the forms as indicated by the accumulation of droppings in the bottom. There has been no evidence nor reports that would indicate sage grouse bury themselves in the snow the way ruffed grouse do.

Attempts were made to capture sage grouse on their night roosting forms by shining them with lights. Under no circumstances would they allow a close approach, but would flush while the light was several yards away. They were extremely alert on their roosting grounds. On many occasions I have flushed grouse from the sagebrush at all hours of darkness while driving along wagon trails and unimproved roads. In April, 1949, I mired my pickup truck in the mud and was forced to make an all-night trek on foot to reach a main highway. At this time large numbers of birds were migrating into their breeding territory, and throughout the night small groups of sage grouse were flushed from roosting areas along the path of advance. This behavior would be an important means of avoiding night attack by natural enemies; likewise, it would enable roosting birds to escape the heavy losses inflicted on other upland game birds such as pheasants, quail, and partridge by snow and ice storms.

Attitudes in Walking and Flying

Sage grouse are not well-adapted for running, due to their heavy bodies and short legs. Their most rapid gait might be described as a fast walk sometimes used in approaching strutting grounds, when engaging other males in combat on strutting grounds, or when moving into alfalfa feeding areas. Compared to quail or pheasants, they never display a fast getaway over the ground. When hard pressed, they either take to the air or attempt to take cover. At times they utilize a most deliberate and hesitant speed of advance over the ground. This is especially demonstrated by hens after they have hidden their broods, and by all adult birds upon approaching the entrances of live-traps. Their complete reluctance to fly is best exemplified during live-trapping operations on the strutting grounds; here they are maneuvered into extended wings only 18 inches high, and thence into the funnel-like opening of the trap proper, simply by the movements of a vehicle behind them as they walk along the ground. Young birds with their mother exhibit surprisingly fast running movements when scattering, or after alighting from flight following disturbance. In their normal walking stride the individual footprints almost touch each other, and may be used to identify sage grouse in areas where other upland game birds occur on the same range. Movements to and from daily feeding and watering areas are more likely to be accomplished by walking the entire distance than by flying. At times the birds utilize combinations of both methods. In these latter instances, the overall movements of birds may cover from 1 to 2 miles, although distances of less than 1 mile are normally involved in going to and from alfalfa fields. Birds habitually utilize the same avenues of ground approach to their feeding areas, and this habit is exploited to the utmost in selecting locations for erecting the traps. Regular beaten pathways are created through the vegetation bordering alfalfa fields, and these are easily detected without actually observing the birds' movements.

Sage grouse are not adept at swimming, although adult birds have managed to flap their way to the bank on two occasions in which they were purposely thrown into small streams. Dead juvenile birds are occasionally observed floating down irrigation canals and ditches. Except near irrigation systems, there would be indeed few places necessitating swimming activities.

As stated previously, sage grouse possess strong powers of flight. They do not spring from the ground like pheasants, nor do they burst from cover with the initial speed of ruffed grouse. Their large size and short legs force them into a laborious take-off involving the full use of their wings. Their legs provide very little spring effect in getting off the ground. Adult cocks have much more difficulty than hens in initiating flight. Once airborne grouse have leveled off, their flight rivals that of the fastest game birds, and can be accomplished at altitudes of several hundred feet for long distances. Large flocks of sage grouse may be observed in migratory flight during the spring and fall seasons at altitudes ranging from 200 to 500 feet. Normal flight is made at altitudes under 100 feet and is not erratic.

Girard (op. cit.) believed that sage grouse are capable of making longer sustained flights in winter than in the summer because they do not become fatigued so readily at low temperatures. In late fall he observed flights of 5 or 6 miles for both cocks and hens, and has recorded average flight speeds of about 45 m.p.h. for hens and 49 m.p.h. for cocks. The highest rate of speed for either sex approached 70 m.p.h.

Health and Sanitation

On their native sagebrush ranges, sage grouse seldom have attained population densities sufficiently high to result in unhealthy or unsanitary conditions, since their range has been practically unlimited. However, with the construction of small reclamation projects (20,000 acres or less) and with the increased agricultural development in the midst of extensive sagebrush areas, sage grouse populations in the vicinity of these projects responded by concentrating to abnormally high levels of abundance in alfalfa fields during the dry summer season. Such conditions of overcrowding may actually set the stage for disease outbreaks. In alfalfa fields in Eden Valley the soil and water are readily contaminated from repeated daily use by literally thousands of birds. Contamination of this nature is believed to have been largely responsible for the localized epizootics invariably associated with sage grouse in irrigated alfalfa fields during the past several years.

The birds are clean in their daily habits. Dust baths are taken by sage grouse of all ages. The sand and loam soil types furnish ideal facilities for dusting activities which are performed generally during the midday resting period. Dirt roadways are the scenes of much dusting, and undoubtedly lure birds into these locations. Sage grouse have never been observed to bathe in open water.

During periods of activity defecation follows no set pattern of deposition. However, when the birds are on roosting grounds, the droppings accumulate to form the characteristic roost-pile, composed of several dozen individual droppings. These are the so-called "intestinal droppings" which originate in the intestinal tract, proper. They are chalky brown to greenish in color, exhibit a solid texture, and contain items of fibrous undigested food. Sage grouse also deposit "caecal droppings," which are formed in the two caecal ducts or blind pouches at the juncture of the small and large intestine. Caecal droppings have a paste-like, watery texture and a dark brownish-black coloration. Upon ejection, the caecal dropping forms a small pool which rapidly dries, becoming solidified within a few days. The regularity of elimination of caecal droppings was undetermined, although a few are apparently passed daily, since one or two inevitably are present in the roosting forms, and they are scattered liberally over strutting grounds. They vary in size from droplets to blotches as large as 3 inches in diameter.

Another type of fecal material known as the "clocker dropping" is deposited by the hen shortly after she departs from the nest on daily excursions for food and water. It represents an accumulation of fecal material in the cloaca while the female is incubating. It is discharged in one complete mass which may be several inches in length and an inch or so in diameter. If the hen walks from the nest site it is usually found within a few feet. However, if she is flushed from the nest it is dropped while in flight, and may be found several hundred yards from the nest. In any event, it is a good indicator of nesting activity and quite often can be used as an aid in locating the actual nest site.

Flocking

Except during the incubating period, sage grouse seldom are encountered singly in the field. The predominant social order is the flock. During the major portion of the year the birds segregate themselves into flocks according to sex, although many such groups contained a few individuals of the opposite sex.

Group behavior aids in the survival of individual birds in the sense that the entire flock generally exhibits synchronous responses to the influences threatening to decimate their numbers, whether these be weather, natural enemies, or man. On the other hand, due to their numbers and wide dispersion, a group of birds is more easily detected than would be a single bird. Daily activities of the birds within the flock organization are well-coordinated, as witness the large numbers of birds which can be live-trapped, often within a period of a few minutes.

Following destruction and desertion of their nests, individual hens gradually assemble into groups of so-called "dry hens," which by early June may number several dozen birds. As many as forty-five of these hens have been seen in a single flock. As the summer wanes, it is presumed that they are joined by other hens which have successfully reared their young. During the winter months, immature birds of both sexes are often associated with the mature hens. During October and November, birds of both sexes and age groups assemble in large flocks preparatory to their movement to winter ranges. Similarly, large groups of birds intermingle in closely associated flocks in early spring as they move onto the breeding grounds. This is particularly noticeable following a severe winter accompanied by a heavy snowfall over the breeding range. On these occasions, the birds tend to concentrate along the periphery of the receding belt of deep snow, awaiting the melting of the snow and the opening of their breeding areas. On March 29, 1950, I witnessed such a sight, as thousands of sage grouse concentrated on the lower drainages of the Dry Sandy—Pacific Creek country en route to their summer ranges which were then still blanketed under several feet of snow. For a distance of 15 miles along the Lander-Farson highway northeast of Eden Valley, large flocks of birds were continuously in sight. They seemed to be literally spread over the entire expanse of sagebrush visible from the highway.

Migrations and Seasonal Movements

The seasonal movements of sage grouse populations have often been a source of much speculation and mystery. Generalized patterns of seasonal movements have been recognized by many observers. There has been little or no definition of the character and magnitude of these movements, nor of flock composition. A correlation of these migrations with climatic changes has also been lacking. Prior to this work, no studies involving large-scale experimental releases of banded birds had been made either as an aid in understanding sage grouse migrations, or in evaluating movements and dispersal following transplanting operations.

Girard (op. cit.) stated that sage grouse do not migrate, although he mentioned that due to snow conditions they usually spend the winter at lower altitudes than they frequent in the summer. According to Batterson and Morse (op. cit.) sage grouse are semimigratory. On some areas observed by the latter workers in Oregon, seasonal population shifts involved a summer movement down to the valleys and agricultural areas, a late fall movement to hills above the breeding and nesting areas, and a late winter movement back to the strutting and nesting areas. Sage grouse in Harney County in southeastern Oregon nested at low elevations, and moved upward in the summer to elevations of 8,000 feet. During the fall they migrated downward to winter at lower elevations.

According to South Dakota Game Warden Harry Henderson (1951) sage grouse annually migrate into Harding and Butte counties, South Dakota, from breeding areas in Wyoming and Montana. In severe winters large numbers of birds are involved in these movements, whereas in mild winters few birds are encountered. During the winter of 1949 Henderson counted about six hundred sage grouse on the Owl Creek flats a few miles east of the Montana line. He believed that the birds drifted into South Dakota for purposes of feeding since at the time there was considerably more snow in Montana than in South Dakota.

For many years Olaus Murie has observed migrations of sage grouse out of the Jackson Hole country into Idaho via the Snake River Canyon. Honess (1941) made a brief study of sage grouse wintering areas and migrations in the Jackson Hole region. He reported that the number of sage chickens wintering in the area was related to snow depths, which, of course, regulated the amount of exposed sagebrush. Near Porcupine Creek, Honess observed a flock of forty-three birds in full flight down the Snake

River, and when last observed, they were well into the canyon below the mouth of the Hoback River. More than two hundred additional birds were observed to be wintering on scattered areas of exposed sagebrush in Jackson Hole.

A juvenile male (No. 2666) was trapped in Eden Valley and released with ninety-one other birds near the Hoback Pass, about 90 miles northwest of their point of capture on October 1, 1949. This bird was killed by an Idaho hunter approximately 25 miles northeast of Soda Springs, Idaho, on September 17, 1950. In this case, then, a bird was recovered some 75 airline miles and two high mountain ranges westward from its point of release. Movement was presumably down the Hoback River to the Snake River and thence into Idaho, or else down the Green River and onto a wintering ground in the vicinity of southwestern Wyoming, northeastern Utah, or southeastern Idaho which was shared commonly by Idaho and Wyoming birds. These migration records emphasize the interstate aspects of sage grouse movements.

Seasonal Movements of Sage Grouse in the Green River Basin

The winter of 1948-49 was considered to be one of the most severe in the history of Wyoming. In Sweetwater County, January, 1949, was the coldest month recorded in more than twenty years. Snowfall was above average, and due to a prolonged period of high winds and ground blizzards the snow blew and drifted extensively. Storm conditions existed almost continuously from late December until mid-February.

By early December, 1948, the majority of the sage grouse in the Eden Valley—Pacific Creek region had withdrawn from their normal summering ranges. During late January and early February, flocks of several thousand birds of both sexes were observed in an area of unusually heavy sagebrush from 10 to 25 miles south of Eden Valley. All local residents commented on the large numbers of sage grouse observed during the height of the 1948-49 blizzard along U. S. Route 187, between the Wells shearing pens and 14 Mile Hill north of Rock Springs. While on aerial reconnaissance over the Little Colorado Desert in late January, Deputy Game Warden Boyd Charter located a flock of sage grouse about 20 miles west of Farson, which extended for several hundred yards on either side of his plane, and contained estimated thousands of birds. He stated that in his many years of field experience in sage grouse range, this concentration of birds was the largest he had ever encountered.

Extensive field work in the Eden Valley—Pacific Creek—Big Sandy area during February and March, 1949, revealed a few widely scattered flocks numbering from ten to seventy birds. These flocks of predominantly adult males were wintering in areas heavily utilized by breeding populations during the summer. No significant population changes nor major movements of birds occurred in this region prior to mid-March. By the third week in March, the heavy snow cover had disappeared in Eden Valley. Many flocks ranging from 25 to 200 birds of both sexes appeared in the upper Pacific Creek region during the first week of April, 1949. In Eden Valley, large numbers of females were encountered between April 7 and April 11 in areas where previously the birds had been scarce. In 1949, both sexes seemed to exhibit a distinct movement onto breeding and nesting territories in early April, although more intensive strutting ground studies in succeeding years indicated that mature males ordinarily preceded the females into breeding areas by several weeks.

Extremely mild and dry weather was experienced in western Wyoming during the fall of 1949. On the study areas two blizzards of short duration early in October provided an impetus for sage grouse and antelope to concentrate. By November, however, a period of extended warm, dry weather resulted in a dispersion of both sage grouse and antelope from concentration areas along Dry Sandy Creek. In Sweetwater County, October and November, 1949, were among the warmest and driest fall months on record. Average daily maximum and minimum temperatures for October at Farson were 40 degrees F. and 21 degrees F., respectively. Precipitation in October was 1.98 inches, mostly recorded during the aforementioned storms. Daily maximum and minimum temperatures in November were 50 degrees F., and 14.5 degrees F., respectively. Precipitation during November was .86 inches, recorded during a snowstorm on November 9 and 10.

During the period from October 31 to December 12, 1949, roadside censuses were conducted throughout the study areas as well as on the Little Colorado and Red Desert areas adjacent to Eden Valley, in order to observe fall movements, flock composition, daily behavior patterns, and predator relationships.

During the fall of 1949 sage grouse again concentrated on the lower drainages of Dry Sandy and Pacific creeks and on the Little Colorado Desert west of U. S. Highway 187, between Eden and 14 Mile Hill. These same areas were focal points for large concentrations of sage grouse during the fall of 1948. There was no complete segregation of sexes or age groups in any of the large concentration areas, although flocks composed predominantly of males or females were commonly seen within the scope of the concentration unit. In almost all cases, large assemblies were composed of loosely-knit flocks whose numbers ranged from a dozen to a hundred birds. Flocks containing from one hundred to two hundred birds were occasionally seen, and one compact group contained an estimated five hundred immature and adult birds. In addition to the birds found in areas of high concentrations, small flocks of sage grouse were distributed throughout the lower Dry Sandy—Pacific Creek drainages and over most parts of the Little Colorado Desert west and south of Farson. Similar patterns of movement were encountered on other drainages of the Basin as flocks of birds moved out of foothill summer ranges onto desert wintering ranges.

Following the first fall storms and the onset of freezing temperatures in October, sage grouse flocks in Eden Valley ceased their feeding activities in the alfalfa fields and dispersed to adjoining sagebrush areas preparatory to movement to lower elevations. During the mild weather of November and early December, individual birds occasionally fed on the basal leaves of alfalfa plants in fields in Eden Valley.

The areas most heavily utilized by sage grouse in the fall also furnished autumn range for hundreds of antelope. Many observations of sage grouse flocks were obtained as a result of seeing birds flushed by herds of running antelope.

For the most part the observations on sage grouse movements, flock composition, and flock behavior during the fall of 1949 were based upon eleven censuses covering a total of 483 miles along roads and trails through the sagebrush. On these censuses, fifty-six different flock units were seen involving about 3,300 birds.

As far as could be determined, a complete and extended movement out of normal summer range had not occurred during the October-December period. However, birds were conspicuously absent from their summer ranges on the extreme upper drainages adjacent to the timbered foothills of the mountains. Movements, resulting in concentrations previously described, had taken place involving a distance of from 10 to 20 miles down the drainages. From the standpoint of weather, winter conditions had not yet set in on the study areas as of mid-December.

The winter of 1949-50 in the Green River Basin was also characterized by deep snows which by early February extended from the foothills well onto the desert regions. A January chinook followed by sub-zero weather was responsible for creating a heavy, impenetrable crust on top of the otherwise deep snows. Eden Valley generally marked the southern boundary of this deep snow belt which stretched north and northeastward to the mountains some 50 to 100 miles distant. Practically all sage grouse flocks were forced to migrate to regions of exposed sagebrush south of Eden Valley and along the lower drainages of the Green River. The snow crust was so hard that mule deer, elk, and moose which normally wintered on the exposed slopes in the foothill country drifted onto the desert regions in large numbers. Several herds of deer were observed from February 1 to 4 as they migrated from the Pinedale-Boulder area into the Eden Valley district. Snow conditions necessitated their traveling on the highway by night. During the daytime they moved a few hundred yards off the right of way, awaiting nightfall and an opportunity to travel again. The entire route was over sagebrush completely covered with crusted snow. Although these deer broke through the hardened snow when walking, it was of sufficient depth and hardness to seriously interfere with their feeding activities. The Eden Valley district supported several hundred mule deer and almost fifteen hundred antelope during the winter of 1949-50. Snow conditions that year were more severe than during the previous winter, although temperatures were substantially higher and ground blizzards infrequent.

In contrast to the two previous winters, there were essentially no major movements of game populations onto desert ranges during the winter of 1950-51. It was perhaps one of the mildest winters on record, and over much of the desert and foothill region the ground remained bare of snow throughout the winter. Large flocks of sage grouse and herds of antelope which would have normally migrated into areas south and west of Eden Valley remained in regions essentially classified as summer range. There was, of course, a limited movement out of the ranges adjacent to the mountains, since these areas always receive a fairly heavy snowfall in the winter.

In summary, it may be stated that during the fall season sage grouse of both sexes and all age groups assembled in flocks numbering from several hundred to thousands of birds preparatory to movement to their respective winter ranges. Observations on sage grouse movements for three years in the upper Green River Basin disclosed that these large groups of birds habitually assembled and remained in the lower regions of their spring and summer territories until the arrival of severe snowstorms and winter weather. As heavy snow and ground blizzards swept over the sagebrush, reducing the availability of their chief source of winter food and shelter, sage grouse flocks flew to regions of exposed sagebrush at lower elevations. These flights not only involved altitudinal movements, but consisted in most years of distinct migrations to areas from 50 to 100 miles distant from their normal summer ranges. These zones of winter flock concentrations were removed from deep snow belts, and were invariably located on the semiarid desert ranges which, due to the lack of spring and summer water supplies, supported relatively small numbers of breeding birds. Movements of sage grouse onto their winter ranges were correlated with snow conditions and the availability of sagebrush. In the absence of open water, snow furnished the moisture requirements for sage grouse in the winter. There was a pronounced tendency for adult birds to segregate themselves according to sex on their wintering grounds. Adult males usually wintered in areas of deeper snow and at higher elevations than did the immatures and females. These wintering flocks of males utilized the sagebrush on exposed, wind-swept ridges and buttes for their food requirements.

Postbreeding Movements of Sage Grouse

As was pointed out earlier, sage grouse inhabiting ranges adjacent to agricultural areas customarily move into the close vicinity of the alfalfa, clover, pea, and potato fields at the conclusion of their breeding activities. In the case of males and hens without broods, these movements begin about the first week in June in Eden Valley. Records of population densities and hunting season kill indicate that this movement of adult birds is confined almost entirely to the sagebrush habitat zone lying from 1 to 5 miles from the fields. There is little evidence that birds flock into the valley from distances up to 20 to 30 miles as popularly believed by ranchers. This local theory of movement has evolved as a result of the inability of the local populace to visualize the high capacity of the immediate area around Eden Valley to produce sage grouse. As the summer progresses, the hens and their broods gradually assemble in the vicinity of the ranches and follow feeding patterns similar to the other birds. Segregation of the sexes continues to be the general rule, even in the vicinity of agricultural areas. Live-trapping operations are highly selective in the capture of either males or females depending upon the locations in which traps are placed.

On areas distant from agricultural operations the mature cocks and "dry hens" move to higher elevations following the strutting and nesting period. Various workers have suggested that this movement is correlated with the increased availability of green feed and water at the higher elevations in the summer and early fall; moreover, this seems to be the proper explanation. A local livestock operator informed me that in August, 1921, he observed a flock of several dozen mature cocks above timber line in the Wind River Range at an elevation exceeding 10,500 feet. Sage grouse are encountered in the summer months in the sagebrush parks on top of the Big Horn Mountains of northcentral Wyoming. On many occasions during this study flocks of mature sage grouse have been observed summering in the South Pass area northeast of Farson in the sagebrush areas bordering the lodgepole pine (*Pinus contorta*) forest at elevations from 8,000 to 8,500 feet. On the Dry Sandy—Pacific Creek area adult cocks were found to be relatively scarce in summer, although during the spring season, strutting grounds were well-distributed throughout the entire region.

Tolerances

Environmental

In primitive times, sage grouse inhabited a range comprised of dominant sagebrush and subdominant grass types. Wet, wild hay meadows were abundantly interspersed throughout this range, particularly near the mountain foothills, and were an equally important component of the birds' native habitat. Much of man's early attempts at settlement and ranching were centered around these wild hay meadows. To a large degree, these early homesteading activities simply converted wild meadows into cultivated domestic hay fields with little disturbance of the original habitat as far as the needs of sage grouse were concerned. Ranches were fairly well-scattered and separated by vast tracts of sagebrush. Human population densities in these areas were noticeably light. Sage grouse were extremely tolerant to this type of development which, under good management, changed their environment very little, at least locally. In subsequent years, increased cultivation placed heavier demands upon water for irrigation purposes, and in a large measure diverted water for local uses to the exclusion of downstream areas. Many of the smaller streams around the fringes of the Green River Basin are now used in this manner with the result that water flow downstream from foothill ranches is greatly diminished after the spring runoff. In the selection of nesting areas, hens generally avoid lands adjacent to streams which become dry in early summer.

The Eden Valley irrigation district, although sponsored by the federal government, is at present a small-scale development very similar in nature and scope to the homesteads which have existed for several decades at higher elevations in the Basin. The land area under cultivation (about 9,000 acres) is well-interspersed with native vegetation and lies at the juncture of three main tributaries of the Green River: Pacific, Little Sandy, and Big Sandy creeks. With the advent of the irrigation system in 1908, sage grouse populations presumably remained abundant. However, the factors generally associated with agricultural development on semiarid lands have periodically operated to cause fluctuations in grouse numbers. The most serious decimating factors have been the destruction of habitat from accumulations of alkali, overgrazing, sage grouse control programs (live-trapping and hunting), and disease.

With minor exceptions, there have been few permanent environmental changes of a nature detrimental to sage grouse in the upper Green River Basin. The changes in land use which have occurred have not materially altered the vast habitat available for seasonal occupancy and use by the birds. On other major land units within the realm of sage grouse, the species has not fared so well. At the extreme periphery of their range, sage grouse were exterminated in New Mexico and British Columbia. In other areas, they have been drastically reduced in numbers through exploitation of their habitat and of the birds themselves. On the Salt River drainage of western Wyoming intensive agricultural development has essentially eliminated sagebrush habitat and exterminated sage grouse populations in a mountain valley more than 40 miles long. Similar factors have operated on sage grouse in Jackson Hole, although an isolated population of probably less than five hundred birds remains on a few thousand acres of federally-owned lands in the northern portion of the valley. On areas such as the large government-sponsored reclamation projects on the major river systems of the West, sage grouse have been either reduced to low levels of abundance, or exterminated completely as solid blocks of several hundred thousand acres of sagebrush lands have been converted in agricultural uses. As a result of agriculture and industrial developments, sage grouse habitat has been destroyed along much of the Wasatch Front for distances of almost 100 miles north and south of Salt Lake City. In the face of changing land-use patterns, sage grouse have continued to be specialized and specific in their habitat requirements, especially from the food standpoint. They have not adjusted, and doubtlessly will not adjust, their life processes to fit a pattern of land use which eliminates or seriously disturbs large tracts of the sagebrush-grass type on any of their seasonal ranges.

Between Species

In this study, sage grouse have been observed in close association with ring-necked pheasants, mourning doves, and certain species of waterfowl. There are sage grouse areas in Wyoming which also support considerable numbers of ruffed grouse, sharp-tailed grouse, Hungarian partridge, chukar partridge, or Merriam's wild turkey on the same or adjacent ranges. Sage grouse, as individuals, apparently exhibit no intolerances to the presence of any of these game birds on their range.

From a series of small transplants in the early 1940's pheasants have become locally established on cultivated areas within the Eden Valley irrigation district. Crowing-count censuses conducted during May of 1949 and 1950, revealed a minimum cock pheasant population of six birds per square mile of cultivated land. Actually, pheasants were not distributed evenly over the district. On the Farson segment they were confined to the vicinity of Little Sandy Creek and the main canal; on the Eden or southern segment pheasants were restricted solely to the area bordering the main canal. These were the only sectors which furnished a combination of grain fields and brushy cover, necessary components of pheasant habitat. The crowing and nesting areas of pheasants in Eden Valley seldom extended into sagebrush types. In the summer months large numbers of sage grouse invaded the habitat normally frequented by the pheasants. No direct conflict between the two species was ever noted. Contention during the nesting season would have been improbable since the birds' breeding and nesting areas were located on distinctly different land types.

In regions opened to pheasant hunting, a large percentage of the pheasants quickly disperse into the adjacent sagebrush to escape the gunning pressure. Such behavior has been observed on the Riverton reclamation area in Wyoming and is believed to involve a considerable portion of the population, particularly of cock birds. During the latter phases of the hunting season many hunters take to the sagebrush for their ringnecks. Under these conditions, shooting hours assume great importance in determining whether the birds will or will not be able to move into cultivated areas for feeding. On the Riverton area sage grouse have been largely replaced by pheasants due to the size and intensity of the reclamation program.

Various species of waterfowl share habitat with sage grouse in the Green River Basin. Previous mention has been made of the mallard and pintail duck nests which were found on the Pacific Creek nesting plots in close proximity to sage chicken nests and on actual sites utilized by the grouse. Green-winged teal, gadwalls, shovellers, and baldpates also frequented aquatic habitats in the midst of sagebrush range types during the spring, summer, and fall seasons. Several pairs of Canada geese annually nested along stream courses in Eden Valley. The entire upper Green River system produced an abundance of ducks and geese, many of which were reared on habitat utilized also by sage grouse.

Mourning doves, a nongame species in Wyoming, are encountered in varying numbers over practically all the semiarid desert portions of the Basin. They are most abundant around agricultural settlements and along main waterways.

Density Limits of Mixed Populations

In his field studies of the effects of grasshopper insecticides upon game bird survival in areas of northcentral Wyoming, Post (1950b) presented data on the numbers and kinds of game birds encountered on the study plots. On a relative basis these data give an indication of the density limits attained by sage grouse in certain areas containing mixed game bird populations. Post established a series of approximately 60-acre sampling units which were thoroughly covered by field crews at weekly intervals throughout the summer. The figures presented in Table 24 represent the average number of birds observed by field crews per day on plots varying from 50 to 75 acres. In addition to the species listed, chukar partridges are established on several of the areas studied, but were not observed on the plots.

It may be concluded that sage grouse tolerate practically every species of game bird known to exist on their range in Wyoming.

TABLE 24

Average Number of Game Birds Seen Per Day on Approximately 60-Acre Plots in Sheridan and Johnson Counties, Wyoming—1950
(After Post, 1950b)

Location	Sage Grouse	Pheasant	Hungar'n Partridge	Sharptail Grouse	Gambel Quail
Red Fork, Powder River	0	2	0	0	0
Middle Fork, Powder River	16	4	0	0	3 (b)
North Fork, Powder River	8	3	15	0	0
Control—Powder River	24	17	1	0	0
Towne Draw	9	0	0	8	0
French Creek	0	9 (a)	4 (a)	0	0
Crazy Woman Creek	41	6	7	0	0
Burnet Creek	116	10	0	0	0
Beaver Creek	15	2	0	0	0
Control—Buffalo	16	2 (a)	9	0	0

(a) Maximum number seen per day
(b) Total observed during the summer

Minimum Units of Range and Population

In as much as the sage grouse is a wide-ranging species, it is probable that the minimum unit of range required for continued grouse survival will be large. The truth of this speculation has been demonstrated in the Green River Basin where the seasonal ranges of individual sage grouse packs may encompass several thousand square miles and may extend from foothills in the upper limits of the Basin to winter ranges 50 or 100 miles down the drainage.

Although the minimum unit of range for sage grouse is high, there is evidence that the minimum population unit is low. For instance, sage grouse were completely exterminated in New Mexico about 1900. Several plantings were made on former sage grouse range in 1933 with birds obtained from South Dakota and Wyoming. From these initially small reintroductions, the birds have survived and spread somewhat. New Mexico game officials now predict that sage grouse will eventually re-establish themselves in that state.

Small populations of birds, partially isolated by mountain barriers or agricultural development, occur at scattered locations throughout the species' range. Jackson Hole, Wyoming, supports a residual population of about five hundred birds. In Utah, particularly, there are many discontinuous flocks, each estimated to contain from 50 to 500 birds. In other states, small localized populations are actually a component of much larger concentrations in adjoining states. The island of habitat in southwestern North Dakota, containing less than fifteen hundred birds, is merely an extension of range situated in northwestern South Dakota and southeastern Montana.

The extreme mobility of the species, as well as the nature of their courtship and breeding activities, practically assures that the entire population, no matter how small or scattered, will exercise its maximum efficiency and highest potential in reproduction. Such behavior has been of great value in the restoration of depleted sage grouse populations in many parts of the birds' range, as well as in the maintenance of remnant populations adjacent to areas of intensive land use.

Domestication

There have been very few organized attempts at rearing sage grouse in captivity. Numerous individuals, out of curiosity, have conducted artificial propagation experiments by placing sage grouse eggs under domestic hens. Hatching is usually successful, but the young grouse chicks exhibit none of the tolerances for survival in captivity that are displayed by artificially reared pheasants or quail. Unless placed in an enclosure, sage grouse young disperse soon after hatching and show no affinities for the foster parent.

Batterson and Morse (op. cit.) reported the results of one small experiment with artificial propagation, which they conducted in Oregon in 1942. Seven chicks from a clutch of nine eggs were hatched by a bantam hen. By utilizing a selective diet (Table 25) and confining the hen and chicks in a runway with a slated coop, six chicks were successfully raised to the age of six weeks and then liberated.

TABLE 25

*Percentage Utilization of Foods by Six Sage Grouse Artificially
Reared in Oregon (After Batterson & Morse)*

		Percent of Diet					
Food	Age in Weeks	1	2	3	4	5	6
Ant eggs and ants		55	50	50	40	30	20
Lettuce		30	25	25	25	25	20
Alsike clover leaves		----	10	10	5	10	10
Wild mustard leaves		10	5	5	15	15	15
Grass tips		5	----	----	----	----	----
Sagebrush leaves		----	5	5	5	5	5
Grains and seeds		----	5	5	10	10	10
Dandelion		----	----	----	----	5	10
Shepard's purse seeds		----	----	----	----	----	10

In the light of our present knowledge, large-scale propagation
of sage grouse under game farm conditions is not considered
possible. Even if practicable, it would not be justified at the
present time in view of the wide distribution and abundance of
the species. A program of successful artificial propagation would
necessitate years of experimental work at high cost, and would
involve efforts similar to those expended by Arthur A. Allen in
rearing ruffed grouse in New York State. As long as the sage
grouse remains well-distributed and abundant over large sections
of its range, live-trapping of wild birds will provide the best
possible source of planting stock, if conditions warrant a planting program.

Food and Water Requirements

Sagebrush and Grouse Survival

Few animals are as highly selective in their feeding habits as
the sage grouse. During the spring and summer seasons, the diet
of the birds includes a large variety of both plant and animal
food. At other seasons of the year they are predominantly vegetarian, and within this category the plants of the sagebrush
group constitute the major portion of their food supply. In the
winter period, these plants are utilized almost exclusively in preference to other types of food. Their preference for sagebrush

is readily explainable. Since it is an evergreen shrub, it is practically the only plant food available throughout the entire year, regardless of snow conditions. It is evenly distributed in one form or another over millions of acres of western plateaus, prairies, and intermontane valleys. It is highly palatable to sage grouse, and from a nutritive standpoint furnishes a very high quality forage. Few people have appreciated or understood the tremendous importance of sagebrush to the survival of sage grouse populations. Sufficient food habits studies of sage grouse have now been made to emphasize the significance of sagebrush in the diet and to demonstrate the specialized feeding behavior of this bird. The welfare and continued survival of sage grouse populations is more closely influenced by the availability and distribution of sagebrush than by any other factor in the birds' environment.

Food Habits Studies

Cottam (1930) stated that in most localities of the West during the greater part of the year the adult birds feed to a large extent on the leaves and tender shoots of sagebrush. From the rather limited food habits data then available, he believed that young grouse fed to a much greater extent on insects than did the mature birds, although both adults and young at times gorge themselves with ants, grasshoppers, crickets, and beetles.

The studies by Griner (op. cit.) of sage grouse food habits at Strawberry Lake, Utah, conformed closely with the results reported by Girard for Wyoming sage grouse. From analyses of stomachs collected from May to October, Griner discovered that almost 98 per cent of the adult bird's diet was composed of plant material of which 77.5 per cent consisted of two species of sagebrush, *Artemisia tridentata* and *A. cana*. The *Compositae* family contributed 86 per cent of the total plant diet. Grasses were next in importance, making up 4 per cent of the total contents. Slightly over 2 per cent of the adult bird's summer food was animal material, principally ants. Juvenile birds consumed more than 50 per cent plant materials during June and July, increasing to 95 per cent in August and September and to 99 per cent in October. During the first month of their lives, Utah birds apparently developed an appetite for sagebrush, since it amounted to 25 per cent of their diet at this early age. Ants were of equal importance to sagebrush in the June and July food preferences of young birds. Based upon observations of feeding habits and analyses of droppings, Griner concluded that the different species of sagebrush supplied nearly 100 per cent of the sage grouse winter diet.

Food habits of Oregon sage grouse have been briefly summarized by Batterson and Morse (op. cit.), who have stated that during the summer the foods of the bird consist mostly of alfalfa leaves, dandelions, clover, wild mustard, and insects. In the fall and winter the primary food is sagebrush. A variety of sagebrush, weed plants, seeds, and insects comprise the food taken in the spring.

Martin and associates (1951) have presented a comprehensive summary study of food habits material from 212 sage grouse collected throughout the birds' range. Plant material formed approximately 95 per cent of the food consumed at all seasons. The small amount of animal food eaten consisted of a great variety of insects, chiefly ants and beetles (particularly ladybird beetles) and true bugs, such as the chinchbug. Nearly three-fourths of the food examined from these grouse consisted of the leaves and flower clusters of the different species of sagebrush.

Foods of Wyoming Sage Grouse—The Food Habits Section of the U. S. Bureau of Biological Survey analyzed thirty-three stomachs and crops collected by Girard in July and August, 1934, in Sublette County, Wyoming. Plant material represented 88.5 per cent and animal material 11.5 per cent of the total food contents. The *Compositae* family furnished 73 per cent of the vegetable food, mainly represented by the dandelion and six species of sagebrush. In the vicinity of agricultural areas sage grouse were especially fond of dandelions, white clover, and alfalfa. When available during the summer, these plants were consumed in large quantities. Leaves and stems of the grass family were found in all except one stomach, yet composed less than 2 per cent of the total vegetable food. Ants, represented by seven species, constituted almost 10 per cent of the total food contents of the thirty-three stomachs and crops. All material examined contained ants, and 3,600 ants of the genus *Formica* were found in one stomach. Every stomach, except one, held representatives of the *Coccinellidae* or ladybird beetle family. The entire lot of inspected food habits material, estimated to be about 80 per cent of full capacity, contained more than 34,000 ants, 730 other insects, and 4,880 weed seeds.

Wyoming Game Department Studies—In an attempt to contribute additional information on sage grouse food preferences, considerable quantities of food habits material were collected in Wyoming by various personnel of the game department. Seventeen crops and stomachs taken in 1940-41 in the western part of the state were analyzed by the Food Habits Section of the U. S. Fish and Wildlife Service. Sixty-two crops and stomachs were collected in the Eden Valley—Pacific Creek region from 1948 to 1951 and analyzed by this writer. Both of these samples were supplemented with twenty-five stomachs recovered in Sheridan and Johnson counties by George Post, and analyzed by him at the department's research laboratory. In presenting these data on a volumetric percentage basis, the plant and animal foods in this entire lot of food habits material have been lumped together. The major sources of both adult and juvenile sage grouse foods in this material are tabulated in Tables 26 and 27.

The results of these studies conform closely with the findings presented by other workers, and establish reliably that the bird is primarily a vegetarian.

The year-round diet of adult grouse was comprised of nearly 96 per cent plant material, the remainder being animal matter, principally insects. On a seasonal basis, insects assumed their greatest importance in the diets of both adult and juvenile birds in the late spring and early summer. Insects supplied approximately 12 per cent of the foods taken by adult birds in the summer. The rate of insect consumption dropped to about 5 per cent in the fall, and with the exception of a few insect galls which the birds picked from sagebrush, insects formed no part of the winter diet (Table 53).

TABLE 26

Major Sources of Plant and Animal Foods in the Diet of Fifty-nine Adult Sage Grouse in Wyoming—1940-41 and 1948-51

Foods	Volumetric percentage	Number of birds in which found
PLANTS (95.7 per cent)		
Sagebrush (*Artemisia*)	77.1	42
Alfalfa (*Medicago*)	4.3	10
Sweet clover (*Melilotus*)	4.1	4
Dandelion (*Taraxacum*)	4.0	16
Rabbitbrush (*Chrysothamnus*)	2.2	10
Vetches (*Astragalus*)	1.5	2
Salsify (*Tragopogon*)	1.3	7
Grasses (*Poaceae*)	.2	5
Prickly lettuce (*Lactuca*)	.2	4
Clover (*Trifolium*)	.1	3
Pussytoes (*Antennaria*)	.05	1
Unidentified	.6	6
ANIMALS (4.3 per cent)		
Insects:		
Grasshoppers (*Orthoptera*)	3.1	9
Beetles (*Coleoptera*)	.3	23
Insect galls	.6	6
Ants (*Hymenoptera*)	.2	19
Bugs (*Hemiptera*)	.02	7
Other insects	Trace*	2

*Less than .01 per cent

Juvenile sage grouse relied heavily upon insects as food during the first few weeks of their existence. During late May and throughout most of June, insects composed as much as 75 per cent of the juvenile bird's diet, although as the summer waned, the volume of consumed insect material decreased; by fall it represented less than 10 per cent of the total food items selected (Table 54). On a yearly basis, plant material furnished 88 per cent, and animal matter (insects) provided 12 per cent of the immature bird's diet (Table 27).

Plant Foods—Throughout the year, the *Compositae* or aster family bulked largest as a source of both adult and juvenile

TABLE 27

Major Sources of Plant and Animal Foods in the Diet of Forty-five Juvenile Sage Grouse in Wyoming—1940-41 and 1948-51

Foods	Volumetric percentage	Number of birds in which found
PLANTS (88.54 per cent		
Sagebrush (*Artemisia*)	46.57	36
Rabbitbrush (*Chrysothamnus*)	12.84	14
Dandelion (*Taraxacum*)	6.04	6
Sweet clover (*Melilotus*)	5.73	4
Prickly lettuce (*Lactuca*)	4.74	2
Grasses (*Poaceae*)	3.27	8
Alfalfa (*Medicago*)	2.78	2
Peppergrass (*Lepidium*)	1.68	1
Salsify (*Tragopogon*)	1.15	1
Linanthus (*Linanthus*)	.41	1
Gooseberry (*Ribes*)	.15	1
Unidentified	3.04	3
ANIMALS (11.46 per cent)		
Insects:		
Ants (*Hymenoptera*)	5.50	27
Grasshoppers (*Orthoptera*)	3.08	7
Beetles (*Coleoptera*)	1.88	36
Insect galls	.81	4
Moths (*Lepidoptera*)	.10	2
Bugs (*Hemiptera*)	Trace*	1
Other insects	.10	2

*Less than .1 per cent

plant foods, contributing nearly 85 per cent of the plant material eaten by adults and 71 per cent of the vegetable food of juvenile birds. Within this family, plants of the sagebrush group, principally *Artemisia tridentata, A. nova,* and *A. cana,* furnished over 46 per cent of the immature diet and 77 per cent of the adult diet. The *Leguminosae* or legume family, represented mainly by clover, alfalfa, and vetches, provided approximately 10 per cent of the plant foods taken by adult, as well as by juvenile sage grouse.

The extreme importance of sagebrush in fulfilling the dietary requirements of sage grouse is shown in Table 53, which depicts the major food preferences of adult birds on a seasonal basis. Only during the summer months does sagebrush compose less than 80 per cent of the total volume of food consumed. Eden Valley birds fed extensively on alfalfa, dandelions, and a few other plants of cultivated areas, this being largely responsible for the marked deviation from a sagebrush diet during the summer, as shown in Table 53. Otherwise, sagebrush would probably have contributed a significantly higher percentage of the summer food. As emphasized in earlier chapters of this treatise, sage grouse food requirements in the winter months (November through March) are satisfied almost entirely by plants of the sagebrush group. It is unlikely that any other species of bird or mammal is more dependent upon a single food item for their subsistence than is the sage grouse. Many species of small mammals and big game animals utilize large amounts of sagebrush in their seasonal diets, but with the exception perhaps of the pronghorn antelope, none of these animals are as dependent upon sagebrush for their seasonal food requirements as the sage grouse. In this connection Martin and associates (op. cit.) have shown that sagebrush is the second most important woody-plant food for wildlife in the mountain-desert region of western United States, furnishing 71 per cent of the seasonal foods consumed by sage grouse and 51 per cent of the yearly foods of antelope.

Animal Foods—The known animal foods of sage grouse are limited entirely to insects (*Insecta*), although very minute quantities of spiders (*Arthropoda*) and millipedes (*Myriopoda*) may be eaten. Within the insect group, ants, grasshoppers, and beetles bulked largest in the diet. On the basis of volumetric measurement, these insects seldom formed more than 10 per cent of the diet in any seasonal period. An exception to this rule is provided by juvenile grouse during the first few weeks of their life. On a frequency occurrence basis, a few insects were generally present in the crop or stomach contents of most sage grouse throughout the warmer months of the year (April to October). Beetles (mainly *Coccinellidae*, ladybird beetles; *Chrysamelidae*, leaf beetles; *Tenebrionidae*, darkling beetles; and *Carabidae*, carab beetles) occurred in minor quantities in 59 out of 104 stomachs examined from Eden Valley and vicinity. Ants (principally *Formica rufa* and *Tapinoma sessile*) were present in 46 of these stomachs. Most of the grasshoppers (*Melanoplus*) appearing in the material studied, were from grouse collected in the north central part of Wyoming.

Insect galls, presumably caused by certain gall midges (*Itonididae* of the order *Diptera*) and gall wasps (*Cynipidae* of the order *Hymenoptera*) occur rather commonly on sagebrush plants in the Green River Basin. They were consumed in surprising numbers by adult sage grouse during practically all seasons of the year, although on a volumetric basis, galls seldom contributed more than 1 per cent of the total food consumed.

Since the nature of insect galls on plants is not well-understood, even by entomologists, explanations for the utilization of this type of food by sage grouse are purely speculative. Galls have been defined by Brues (1946) as abnormal growths which develop in the tissues of plants, due ordinarily to the introduction of foreign substances. Insects are the most conspicuous causes of galls.

There is a probability that sage grouse select galls for the purpose of obtaining insect eggs or larvae which are developing inside the growth. The complex nature of the gall food materials, which are similar to the proteins, also may be instrumental in their being utilized by the grouse.

In the vicinity of irrigation projects, sage grouse have shown a distinct preference for alfalfa, clover, dandelions, as well as the leaves of potatoes, beans, peas, and other truck crops. In addition to supplying luxuriant forage, many of these areas furnish abundant quantities of insects and continuous sources of drinking water. Large-scale movements into agricultural areas are generally associated with increasing aridity in the sagebrush types as summer progresses. The studies conducted in Eden Valley indicated that habit plays an important part in this seasonal invasion, once it has been established.

The nutritive values of food compounds in alfalfa hay compare favorably with the values of similar foods found in sagebrush. During the severe winter of 1948-49, sage grouse were observed to feed sparingly upon second-cutting alfalfa used for emergency feeding of antelope on winter ranges in Sweetwater County, Wyoming.

There are no records to indicate that sage grouse are susceptible to direct poisoning from any species of plant. However, certain selenium indicator plants such as the vetches are consumed in some quantities by sage grouse in certain areas of Wyoming.

Forage and Digestibility Values of Sagebrush

Samples of fresh sage grouse droppings and sagebrush leaves were analyzed by standard chemical methods at the Wyoming Game and Fish Commission Disease Research Laboratory. Leaves of the black sagebrush (*A. nova*) and sage grouse winter droppings were collected simultaneously from the same site along Dry Sandy Creek northeast of Eden Valley. Field observations disclosed that the birds in this area were feeding predominantly upon black sagebrush leaves. The results of these chemical analyses are presented in Table 28.

The analyses indicated that the droppings contained more crude protein and total nitrogen than did the leaves. The increased total nitrogen seems out of proportion since nitrogen intake and nitrogen output of animals in good health and at average ages should be in balance. There was a possibility of error in the sampling method, since the sagebrush leaves were taken from the plants and not from the crops and gizzards of the grouse.

TABLE 28

*Percentage Composition of Black Sagebrush Leaves and Sage Grouse Droppings**

	Leaves	Droppings
Fat (ether-soluble material)	7.5 %	15.1 %
Crude protein	10.1	14.1
Crude fiber	15.0	21.2
Ash	4.5	5.7
Carbohydrates (nitrogen-free extract)	62.9	43.9
Lignin	7.03	6.55
Total nitrogen	1.61	2.26
Nonprotein nitrogen	0.10	0.34

*Dry weight basis

The increased fat values may be caused by a concentration in the droppings of the nondigestible complex oils from the leaves.

The lignin ratio in forage and feces has been utilized by Cook and Harris (1950) as a means of determining the digestibility of nutrients consumed by domestic sheep. The lignin values determined for sage grouse droppings and sagebrush leaves indicated that the lignin ratio also may be applicable in digestibility studies of sage grouse foods.

The digestibility value of the carbohydrates in black sagebrush leaves was found to be 25 per cent, based upon the lignin ratio technique. Negative digestibility values were apparent for the ether-soluble material (fats) and crude protein due to increased values in the droppings.

Of all plants on the Red Desert which were investigated for food values by the Wyoming Agricultural Experiment Station (1938), the leaves of the big sagebrush (*A. tridentata*) were in a class by themselves. The analyses of these leaves showed 16 to 17 per cent of the total solids as proteins, 14 to 17 per cent as fats, and 46 to 49 per cent as nitrogen-free extract (carbohydrates). The protein content was equal to and the nitrogen-free extract was greater than that found in the best alfalfa meal. The remarkable thing was the high fat content. This was twelve times the fat content of alfalfa meal and five times as much as found in any other of the Red Desert plants which were analyzed. The high fat content of sagebrush leaves makes them a very valuable winter food, since fat has a high capacity for producing heat. In this respect, 16 per cent fats is considered equal to 40 per cent carbohydrates. Analyses of Red Desert plants that remained green throughout the winter disclosed that the chemical composition changed very little from season to season. The forage of the Red Desert had a very high feeding value. After drying, the leaves of the sagebrush had the composition of a concentrate. The leaves of the salt sages also had a nutritive value as high as the best alfalfa hay.

Use of Salt by Sage Grouse
Salt licks have long been recognized as an important component of domestic livestock and big game ranges. There have

been few, if any, recorded instances of upland game birds utilizing salt licks or feeding upon mineralized soil. While engaged in photographic studies in Jackson Hole, I had occasion to closely observe the activities of sage grouse on an area of land which served jointly as a cattle salting ground and a grouse strutting area. A single block of livestock salt, located in the center of the clearing in the sagebrush, was being used by white-faced Herefords en route from local ranches to their summer range allotments on national forest lands rimming the valley. On the morning of May 26, 1951, several sage grouse cocks and hens were observed in courtship activities amidst these cattle. At intervals during the morning's activities, sage grouse were observed to congregate around the block of salt, alternately eating the soil at the base of the block and pecking at the salt block. Similar behavior was noted at another salting ground in the general area. Four strutting grounds in the Jackson Hole area, now included within the recently enlarged Grand Teton National Park, are located on active cattle salting grounds.

In the Eden Valley area and adjacent desert regions, range livestock operations are concerned primarily with sheep. There is little need to salt sheep on ranges which are interspersed with the salt desert-shrub vegetation, since the animals apparently consume adequate quantities from salt shrubs such as *Atriplex nuttallii* and *A. confertifolia.* On the foothill ranges and on the national forests, salting of range sheep is practiced, but the operation differs from that practiced with cattle, in that the salt is distributed in the granular form and not at regularly established locations.

It may be possible that sage grouse require nominal amounts of salt in their diet. In mountainous regions such as the Jackson Hole area and the extreme upper limits of the Green River Basin—both of which contain solid stands of sagebrush to the total exclusion of the salt shrubs—livestock salting operations may supply dietary elements otherwise lacking. If this is true, it would indicate that sage grouse, like domestic sheep, obtain their mineral requirements on the semiarid desert ranges either from the water or the vegetation. There is a need for additional studies on all phases of this question.

Water Requirements

Under the conditions existing over most of their range, sage grouse utilize moisture in one form or another at regular intervals. If free-water is available, they normally make a morning and evening visit for purposes of drinking. Birds have been regularly observed in areas which prevent them from obtaining water from open springs, streams, or ponds. Broods in small numbers are hatched and reared throughout the central arid portion of the Little Colorado Desert west of Eden Valley. These young birds could not possibly reach open water until several days after hatching. Since moisture in the form of dew and rain water is not regularly available on desert ranges, their moisture requirements may be fulfilled partially through the medium of metabolic processes within the bird itself.

The importance of snow in satisfying the moisture requirements of sage grouse has been established on several occasions. In November, 1949, a flock of sage grouse, a few miles northeast of Farson, was photographed while eating snow. Warm weather had practically eliminated an early autumn snowfall, and these birds were congregated around a remnant snowbank on a north slope. Tracks and droppings indicated that they had been utilizing this source of moisture for several days.

Sage grouse have also been observed to regularly visit partially frozen streams in Eden Valley during the late fall months in order to drink through holes in the ice. Many ranchers relate stories of sage grouse flocks coming into their ranch yards and drinking from livestock watering troughs. Although sage grouse apparently do not require open water for their day to day survival, they do utilize it in all cases when available, and attain their highest population densities in those areas which contain abundant and well-distributed surface water supplies. Sage grouse drink in the manner of domestic fowl.

PART III

———

SAGE GROUSE AND MAN

TRAPPING AND TRANSPLANTING SAGE GROUSE IN WYOMING

History of the Program

One of the primary objectives of the sage grouse survey as constituted in 1939 was to accomplish the redistribution and restoration of the species in suitable habitat in Wyoming. Before work was initiated on actual trapping and transplanting, a preliminary investigation of the status of sage grouse was made. By the fall of 1940, the survey party had sufficient information on the status of sage grouse in the state to commence work on the program of restoration. It was hoped that this could be accomplished by live-trapping sage grouse in areas of heaviest concentrations and transplanting the birds to areas of suitable habitat where their numbers had been depleted. Late summer was deemed the most desirable time of the year to undertake this project since most of the young birds were on their own, and the birds of both sexes and age groups were beginning to flock together.

To the state of New Mexico must go the credit for pioneering in the field of sage chicken live-trapping and transplanting. This development, however, was made possible through timely co-operation on the part of the Game and Fish Commission in Wyoming where the initial experimental work was conducted. Over a period of four years prior to and including 1936, approximately two hundred sage grouse were captured in Wyoming by representatives of the New Mexico Department of Game and Fish, and used for restocking in the latter state. The undertaking was successful in a large part due to the instructions regarding the techniques of trapping by J. Stokely Ligon, then of the U. S. Bureau of Biological Survey, who has described the techniques for trapping and handling sage grouse, and who has presented the details of trap construction. Ligon (1946) considered sage grouse to be the most temperamental and unpredictable of all upland game birds and the most difficult to handle in captivity. The present day methods utilized to trap and handle sage grouse in Wyoming are essentially refinements or modifications of techniques originally presented by Ligon. Most of the trapping has been done in agricultural areas that were overpopulated with sage grouse and from which damage to alfalfa fields had been claimed by ranchers. Trapping has been essentially confined to the Eden Valley sector in Sweetwater County, although a few hundred birds have been taken in the Muddy Creek and East Fork areas of Sublette County. The trapping operations were conducted in order to provide ranchers with some measure of relief from damage, as well as to moderate or eliminate damage claims; to develop and improve trapping, holding, and transplanting techniques; to obtain birds for restocking depleted ranges; and to foster studies on growth, sex ratios, and behavior. All birds were leg-banded prior to release as an aid in studying movements and in evaluating the success of the transplanting program.

Intensive live-trapping operations, conducted each year from 1940 to 1943, resulted in the capture of 2,362 birds. The trapping program was resumed in 1946, and has continued each summer through 1950. From 1946 to 1948, only 609 birds were trapped, due largely to inexperienced personnel and to an apparent reduction in sage grouse numbers and a lack of damage areas. During 1949 and 1950, live-trapping of sage grouse was a major activity from midsummer until late October. An abundance of birds, more experienced personnel, and the utilization of additional trapping equipment resulted in a combined catch of 3,196 birds for the two-year period (Table 29).

TABLE 29

Number of Sage Grouse Trapped in Wyoming
from 1940 to 1951

Year	Dates	Number of Birds	Locality of Capture
1940	Sept.-Nov.	644	Eden Valley—Sweetwater Co.
1941	Aug.-Sept.	262	Eden Valley—Sweetwater Co.
	Sept.-Oct.	344	Oregon Springs—Sublette Co.
	Oct.	48	Saratoga—Carbon Co.
1942	Sept.-Oct.	575	Eden Valley—Sweetwater Co.
			Oregon Springs—Sublette Co.
1943	Sept.-Oct.	489	Muddy Creek—Sublette Co.
1944		No Trapping Conducted	
1945		No Trapping Conducted	
1946	Aug.-Oct.	117	Eden Valley—Sweetwater Co.
1947	Sept.-Oct.	377	Eden Valley—Sweetwater Co.
1948	July-Oct.	115	Eden Valley—Sweetwater Co.
1949	Aug.-Oct.	1,123	Eden Valley—Sweetwater Co.
1950	July-Oct.	2,073	Eden Valley—Sweetwater Co.
1951	May	14	Eden Valley—strutting grounds

The combined catch for the entire period from 1940 to 1951 was 6,181 birds; thirty-four of these birds (cocks) were captured on Eden Valley strutting grounds in 1950 and 1951.

Approximately 5,881 sage grouse have been liberated in nineteen counties of Wyoming during the period from 1940 to 1951 (Table 30). The release localities of these birds are shown in Figure 14. The remaining birds have been utilized for experimental purposes or transferred to the New Mexico game department in exchange for wild turkeys. Casualties accounted for 185 birds, or 3 per cent of the birds trapped.

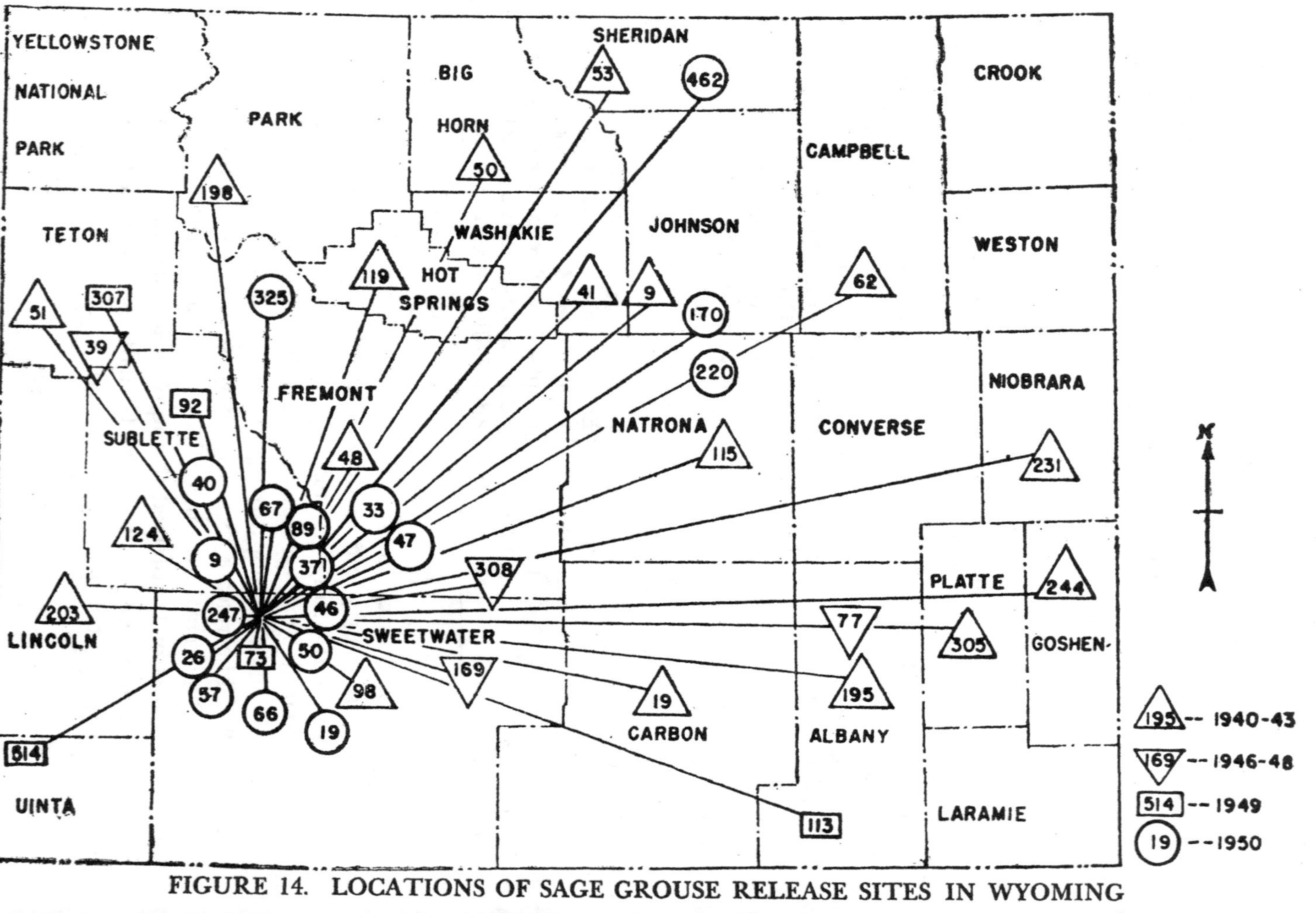

FIGURE 14. LOCATIONS OF SAGE GROUSE RELEASE SITES IN WYOMING

TABLE 30
Release Localities of Sage Grouse Transplanted in Wyoming from 1940 to 1951

County	Year							Total Released
	1940-43	1946	1947	1948	1949	1950	1951	
Albany	195		77		113			385
Big Horn	50							50
Campbell	62							62
Carbon	19							19
Fremont	48	30	219	59		405		761
Goshen	244							244
Hot Springs	119							119
Johnson	9					170		179
Lincoln	203							203
Natrona	115					220		335
Niobrara	231							231
Park	198							198
Platte	305							305
Sheridan	53					462		515
Sublette	124				92	242		458
Sweetwater	98	87	80	2	73	511	14	865
Teton	51			39	307			397
Uinta					514			514
Washakie	41							41
	2,165	117	376	100	1,099	2,010	14	5,881

Trapping Equipment

Except for certain modifications, the trapping equipment used in 1949 and 1950 was essentially the same as developed and used by trapping crews in previous years. In 1949 two traps were used, each identical in pattern and equipped with a set of wings 150 ft. long. The traps were constructed of 2-in. mesh, cotton netting made into a rectangular enclosure 14 ft. wide, 28 ft. long, and 42 in. high. The entire structure was supported by three transverse arches, constructed from $\frac{3}{4}$ in. by 14-ft. iron pipe, each of which was set into a pair of posts fabricated from $1\frac{1}{2}$-in. angle iron. The posts were 52 in. long and pointed at the bottom. A cross piece was welded about 10 in. above the base of the posts for support, as well as to indicate proper depth-setting. The transverse arches fitted into a section of iron pipe 1¼ in. in diameter by 5 in. in length, which was welded ½ in. from the top and on the inside of the iron post. The posts were driven into the ground with the aid of a driver 34 in. long, constructed from 3-in. steel pipe with heavy circular plates welded onto one end.

For strength and stability the netting was faced on its edges with ¼-in. cotton rope. The front and rear ends of the reinforced netting were secured to the arches by means of 1¾-in. metal rings spaced 8 in. apart. Metal rings were also fastened to the rope-facing at each of the four lower corners of the trap. The entire trap was secured to the ground at approximately 3-ft. intervals with iron stakes 8 in. in length by ⅜ in. in diameter, with a short lip of ¼-in. rod welded to the top end for securing the netting. In place of the iron stakes, heavy spikes may be utilized.

The major modification to the traps used in 1949 consisted in eliminating the provisions for camouflaging the entrance. Previous trapping experience had disclosed that sage grouse were hesitant to enter traps whose entrances had been camouflaged with vegetation and featured a style of construction differing from the rest of the trap.

The newly designed entrance panels were of a built-in, permanent type, and featured the same type of netting used in constructing the sides and top of the trap. The entrance panels were the same height as the trap, and were permanently laced to the front edges and top of the trap to a point 6 ft. inside the front arch, to create a *V*-shaped opening. The edges of the entrance panels were also faced with ¼-in. cotton rope. Each panel was fitted with a metal ring at the lower inside corner at the point of entrance into the trap proper. A section of lath was used to create and support an entrance which should be about 10 in. wide at the base. Each trap was provided with two wings of 2-in. mesh, cotton netting, each 150 ft. long and 18 in. high. The wings were supported by permanently attached wooden pegs driven into the ground at 6-ft. intervals.

Trapping Methods

In previous years several techniques were devised to capture sage grouse. Considerable success was realized by trapping birds in the alfalfa fields or around water holes on the desert. In this method the birds were trapped as they walked into the fields or into watering areas. Another method used with fair success consisted of erecting the trap and wings, and using a vehicle to drive the birds inside. This latter method was adaptable for use in either alfalfa fields following the harvest, or on flocks of birds roosting in sagebrush areas. If used in alfalfa fields, the trap and wings were erected prior to the entry of birds into the field. Otherwise, flocks were located in the brush and traps set up in the vicinity, after which the birds were maneuvered with a truck.

In agricultural areas supporting heavy concentrations of sage grouse, the most efficient trapping technique involves the capture of birds in alfalfa fields. With the exception of a few cocks caught on the strutting grounds, all grouse trapped in 1949 and 1950 were captured as they walked into alfalfa fields. The baiting principle was used for trapping sage grouse. Traps were erected on the extreme outer edge of alfalfa fields with the wings radiating from the trap, underneath the fence, and thence into the sagebrush at a widely divergent angle. Two traps were usually set sufficiently close so that the two inner wings met, forming a *W* figure. In this manner, birds walking into a field could be effectively intercepted along a wide front. Preliminary reconnaissance (observation of the birds' time of entry and departure from the field and surveillance of the tracks in the dirt; disclosed the routes of approach used by the birds, and hence the proper location for erecting the traps. For purposes of feeding and watering, it was customary for sage grouse to enter the fields during the early morning twilight period and again in the late afternoon. Early morning trapping was attempted, but was generally a laborious and an unproductive procedure due to the difficulty of predicting the path of the birds' approach, the problem of erecting traps during darkness, and the preferred habit of flocks to fly rather than walk into fields in the morning. Furthermore, morning trapping seemed to disturb and scatter the flocks, thus disrupting the afternoon trapping. Morning trapping was therefore discontinued.

Traps were set up during the late morning period in alfalfa fields adjacent to sagebrush areas known to be harboring large numbers of birds. Movement into the fields usually occurred between 4:00 p.m. and 6:00 p.m. and involved a wave of birds in a relatively short period. Trapping crews observed the progress of activities from trucks parked several hundred feet on either side of the traps. Birds were removed from the traps as soon as the general movement into the wings and traps had ceased, or when birds began drifting out of the entrance in large numbers. On days when the initial wave came into the fields early in the evening it was sometimes possible to make two catches. If birds missed the wings entirely, it was often possible to flush them from the field and into the wings for a new approach.

White-tailed jack rabbits (*Lepus townsendi campanius*), normally abundant around alfalfa fields, were extremely destructive to the traps and wings due to their fondness for chewing holes in the netting. It was, therefore, necessary to lift the trap netting and wings at the completion of each day's trapping operations.

Ideal conditions for trapping occurred on hot, dry days with a minimum of wind. Excellent catches were made, however, on squally days and on days with winds of moderate to fresh force. Trapping operations during August and September, 1949, were usually productive of large catches, and it was a rare occurrence when the traps failed to take some birds. However, by October and the completion of the hay harvest, the birds became more unpredictable in their daily movements. Flock organization within the fields began to break up as the killing frosts hit the alfalfa. The birds were inclined to fly into the fields as fall weather set in with cooler temperatures. It was during this time of the year that an attempt was made with trucks to drive birds into traps set up in the middle of alfalfa fields. For this type of operation the two traps were set in tandem, end to end. In effect, this created one huge trap 56 ft. long. The entrances on both traps were lifted to make an opening straight to the back of the rear trap. The wings were erected in the customary manner on the outer trap. On the only occasion this method of trapping was used in 1949, a large flock of birds was maneuvered with two slow-moving pickup trucks into the wings and thence into the traps. Forty-nine birds were caught in this manner, although many more were pushed into the wings. The drive would have been more successful if the entrance to the rear trap had been erected to form a holding pen into which the birds already trapped could filter. Without this provision, the birds commenced to drift out of the wide entrance almost immediately after being trapped. This necessitated a fast approach with the trucks to seal off the entrance, and consequently flushed all the birds in the wings. After the alfalfa is harvested, making it possible for trucks to be driven over the fields, the drive method of capturing sage grouse has wonderful possibilities. Enormous catches may be made if some provision is made to hold the birds in the rear of the trap during the drive.

On October 6, 1949, a severe blizzard set in and lasted for three days. Snow accumulated to a depth of 20 inches. Following the disappearance of the snow, no large catches were made due to the failure of the birds to follow their normal feeding routines in the fields. Only scattered flocks frequented the fields, and they exhibited no definite feeding pattern. The last catch of birds was made on October 17. Trapping operations for 1949 ceased at that time.

Except when erected inside the fields for the drive type of capture, the traps were set up on the extreme edge of the alfalfa fields. The wings were attached to the outside entrance of each trap and extended into the sagebrush again at a widely divergent angle. The net wings were only 18 inches high, but this was sufficient to divert the sage grouse into the entrance of the trap. In some instances, the birds would go under the wings if they were not secured close to the ground. It was not necessary to cut away sagebrush since the wings functioned properly when set around or between sagebrush clumps. Trapping was more effective if the entrance was set up to embrace a well-established and unobstructed path of approach. The V-type of built-in entrance resulted in the capture of more birds than was obtained with a camouflaged entrance. It did not seem to discourage birds from entering the trap as did the types of entrances previously used.

The small entryway was difficult for the birds to locate once they had entered the trap, and the inner sides of the entrance panels tended to deflect birds away from the opening. Few birds escaped after once entering the trap.

The large traps now used exclusively for trapping sage grouse have several advantages over the small traps previously utilized. Upon entering the large trap, the birds have a great deal of freedom in their movements, and often considerable time elapsed before they were alerted to their capture. They normally commenced feeding on the alfalfa as soon as they entered the trap. When large catches are anticipated, the lower edges of the netting should be well-staked down to prevent a massive number of birds from crowding against the netting and escaping underneath.

Field reconnaissance and interviews with local ranchers indicated that sage grouse were visiting alfalfa fields in large numbers and doing considerable damage to the hay by their feeding and trampling. Strutting ground and nesting censuses conducted during the spring of 1949 and again in 1950 disclosed that approximately twenty-five thousand immature and adult sage grouse were present within the Eden Valley irrigation district during mid-summer. Practically all these birds were in sagebrush areas lying within a zone 1 to 5 miles in depth, peripheral to some 9,000 acres of cultivated land. Slightly over eleven hundred sage grouse were live-trapped from alfalfa fields in the Eden and Farson segments of Eden Valley irrigation district in 1949. Trapping operations were initiated during mid-August.

With the exception of birds involved in the live-trapping and transplanting program, sage grouse received complete protection in 1949. Excessively high sage grouse numbers continued to exist in Eden Valley throughout the summer of 1950. Trapping operations in 1950 commenced in early July, since sizeable flocks were feeding in alfalfa fields at that time. Trapping operations continued until August 27 at which time they were temporarily discontinued due to the opening of a general open season on sage chickens in southwestern Wyoming. Six days of hunting extended over a three weeks' period. This shooting season scattered and dispersed the birds into outlying sagebrush areas, and trapping operations were not resumed until October 4. Trapping was terminated on November 1. A total of 2,073 sage grouse were trapped on ranches in Eden Valley in 1950. Three hundred and twenty-three birds were taken in October following the hunting season.

Three traps identical in design were utilized throughout the summer of 1950, whereas two traps were used in 1949. Two traps were occasionally set up in tandem along the edge of alfalfa fields with the back of the front trap united to the forward portion of the rear trap. In this manner the rear trap functioned as a holding pen, and effectively kept the birds away from the area just inside the entrance to the front trap. The tandem method of erecting the traps was similar to that used in the drive type of capture, except that the entrance panels of both traps were in position. Once the birds were inside a trap erected singly, they had a tendency to mill around the entrance, thus discouraging other birds from entering the trap. The two-trap combination usually resulted in larger catches of birds than would have been accomplished with two traps placed singly.

Handling Sage Grouse in Captivity

Birds were removed from the traps by means of a long handled, lake trout landing net and immediately placed in a standard game bird farm transport crate for movement to the holding sheds. Sexes and age groups were segregated in separate holding crates as they were removed from the traps. This reduced the casualties caused by birds trampling one another, and greatly facilitated banding operations. The crates held ten adult males or twelve adult females and juveniles. Except for purposes of banding, weighing, and measuring, the birds were left in the holding crates until released at the transplant site. If birds were held more than twenty-four hours, alfalfa leaves and water were placed in feeding trays attached to the outside of each crate. Crates with captive birds were stored in a dark, cool, and secluded building prior to movement to the release sites. The practice of placing sage grouse into holding pens and brailing their wings was used extensively in previous years of trapping, but was discontinued in 1949 and 1950 in order to reduce handling, injury, and shock. Every attempt was made to release birds within forty-eight hours after their capture. If properly handled, casualties were negligible up to two days in captivity. Thereafter, the casualty rate increased rapidly. It was necessary to place a thick covering of sawdust on the floors of the crates each time they were used in order to absorb the moisture from body wastes; otherwise, the feathers became soiled and mortality increased. Birds were transported by truck to release sites as rapidly as loads were available, and under no circumstances were they confined more than seventy-two hours. Release sites were selected upon recommendations of the deputy game wardens and were usually remotely located areas which provided an abundance of water and heavy sagebrush for shelter and food. Narrow ravines and restricted valleys were more desirable as release sites than buttes, ridges, and tablelands, since the birds tended to retain their flock organization when released in a restricted topographical area. Birds which were to be released several hundred miles distant were moved immediately after their capture. Sage grouse have been moved as far as three hundred miles in Wyoming with less than a 3 per cent loss for the entire operation.

Classification of Casualties

The greatest mortality sustained by sage grouse during trapping and transplanting operations was caused from head injuries received by the birds jumping forcibly against the tops of the holding crates. Losses from injuries sustained in this manner may be reduced by eliminating the wooden lids and substituting heavy burlap or canvas tops. Trampling and exposure casualties in the crates may be practically eliminated by segregation of the sexes, by separating the adult and juvenile groups, and by keeping a heavy layer of absorbent material on the floor of the crates.

Actual trapping casualties were held to a minimum by careful handling and by preventing birds from crowding into a corner of the trap and piling on top of one another. At times when extremely large catches were made in a single trap, a bird or two would be trampled to death or suffocated. When removing birds from crates or traps, they were grasped by placing one hand on the humerus of both wings and the other on the abdomen to support the bird's weight. Dislocation of the wings was a hazard unless they were locked over the back and the bird was supported from underneath. Grouse of both sexes resisted handling vigorously, and were easily injured if handled carelessly. Trapping and casualty data on over 2,000 sage grouse handled during the summer of 1950 are presented in Table 31. A large percentage of the casualties incident to banding and to confinement in the holding crates were caused from the head and neck injuries sustained in the manner previously described.

Casualties from all sources accounted for slightly less than 3 per cent of all birds handled during 1950 trapping operations. Adult females exhibited higher casualty rates than birds in other categories. This was attributed to their lowered resistance and greater susceptibility to shock. Adult males sustained the smallest number of casualties due to their large size and greater vigor.

TABLE 31
Classification of Sage Grouse Trapping Casualties—1950

Classification of Casualties	Adult Male	Young Male	Adult Female	Young Female	
Trapping operations	0	2	3	3	
Banding operations	0	2	5	5	
Confinement in crates	4	5	9	4	
In transit	2	6	7	4	
Total casualties	6	15	24	16=	61
Escaped uninjured	0	1	0	1=	2
Successfully released	734	299	590	387=	2010
Total trapped	740	315	614	404=	2073

Strutting Ground Trapping

From May 9 to 15, 1950 an attempt was made to develop a method for trapping sage grouse cocks on their strutting grounds. Two experimental systems of trapping were utilized. In one method the standard type trap with wings was erected to intercept the cocks as they walked onto a strutting ground (Figure 15). The other method featured the trap in the reversed position in order to capture the cocks as they withdrew from a strutting ground (Figure 16). In both methods the wings were laid out in a manner similar to that used when trapping birds in the alfalfa fields. In both systems of trapping the cocks readily entered the wings but were reluctant to enter the trap proper. In the former method as soon as one or two cocks entered the trap they became restless and frightened other birds away from the entrance. This type of behavior was attributed to the fact that definite territorial patterns had previously been established on the strutting ground. Trapped birds were unable to reach their particular niche on the courtship ground and consequently fought the netting. Another source of trouble was presented by a cock which arrived on his territory located within the wings and near the trap entrance. He defended this spot vigorously against the approach of other males. Such a pattern of behavior served to drive away other cocks which encroached upon the entrance to the trap.

The capture of cocks upon their withdrawal from the strutting ground seemed to be the more effective trapping method in the spring. It seems improbable that more than a few cocks could be caught at one time by either method, since in 1950 only twenty birds were taken in a week of trapping. Both morning and evening display periods were used for experimental phases of strutting ground trapping. Operations were delayed until mid-May in order not to interfere with breeding activities.

In 1951, experimental studies were continued on the various aspects of strutting ground trapping. Trapping was confined to the evening display period, and in all cases the traps were erected to intercept the cocks as they departed from the grounds. The technique, as ultimately developed, consisted in allowing the cocks to take position on the strutting ground between the outstretched wings. Two standard-sized traps were erected in tandem on the margin of the strutting ground and in the edge of the bordering sagebrush. The entrance panels were omitted completely from the outer trap, although they were erected on the inner trap. About thirty minutes after sunset a pickup truck was driven into position on the extreme outer edges of the wings. This preliminary movement of the truck had a tendency to cause the birds to drift off the grounds in the direction of the trap entrances. The truck was then maneuvered in a zigzag manner

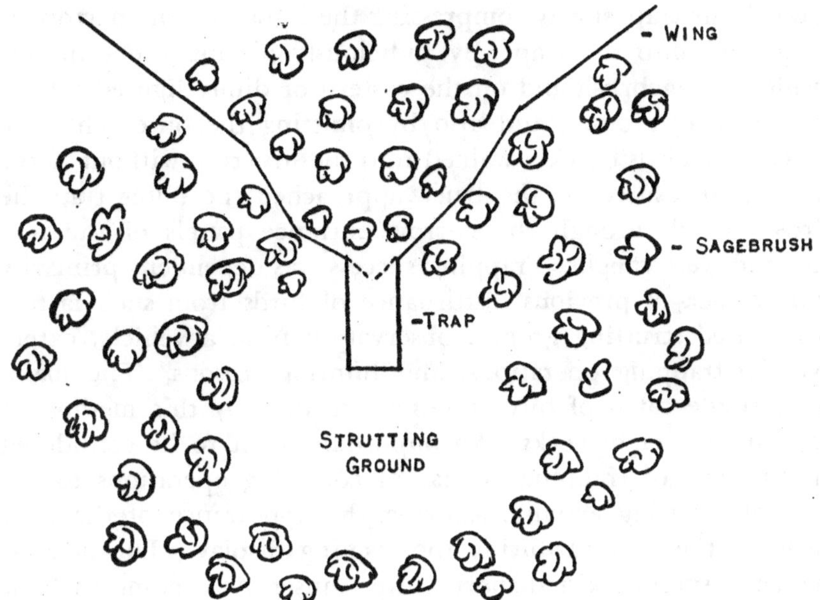

FIGURE 15.   TRAP SET TO CATCH BIRDS APPROACHING
STRUTTING GROUND

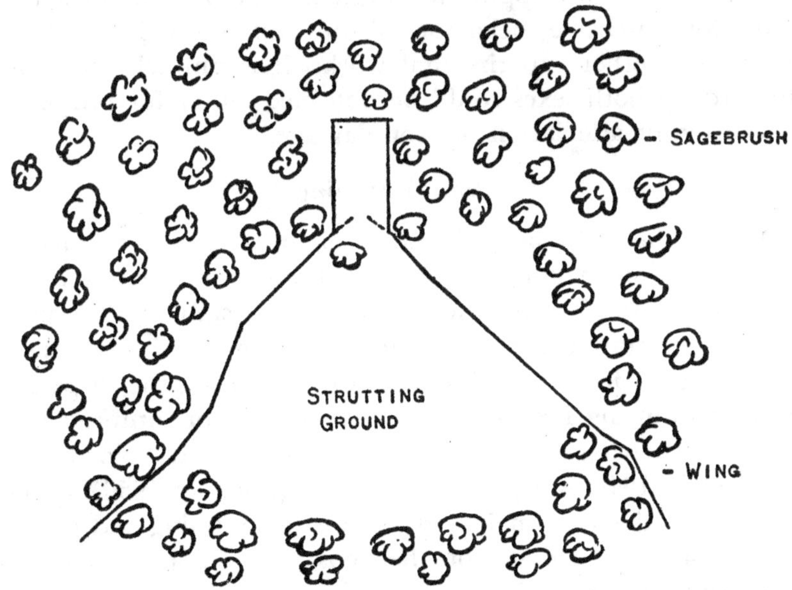

FIGURE 16.   TRAP SET TO CATCH BIRDS WITHDRAWING
FROM STRUTTING GROUND

towards the trap, slowly compressing the birds into the narrowing wings and into the trap. By judiciously placing the trap just inside the sagebrush and on the eastern or dimly lighted side of the strutting ground, and also by omitting the outer entrance panels, the birds quickly walked into the first trap without noticing the enclosure. As the truck approached the outer trap the birds walked through the narrow entrance panels of the rear trap and were caught. Trapping success was dependent primarily upon a lack of previous disturbance of birds from such factors as repeated strutting ground observation from a vehicle, extensive live-trapping operations, and hunting seasons. The maximum single catch of birds attained in 1951 by this method of trapping was eight cocks. An important point to be considered in this type of trapping is that of confining operations to the peak of strutting activity, assuring the appearance of the cock birds on the ground during the evening display. The greatest chance of success existed shortly after sunset. As evening twilight faded the birds were more likely to flush from the grounds than to walk into the surrounding brush.

This method of trapping would mainly be of value in marking birds for intensive behavior studies on individual strutting grounds, or other experimental work. It is unlikely that sufficient birds of both sexes could be obtained either for purposes of restoration or alleviation of crop damage.

Nest Trapping

During the 1949 nesting season eighteen separate hens were live-trapped on their nests, leg-banded, and immediately released at the nest site. Twenty-four females were similarly trapped in 1950. Nest trapping was performed as an aid in tracing the movements of individual broods, in determining the dispersal of females following the nesting period, and in obtaining a record of a hen's nesting activities during ensuing years. No desertion occurred in 1949 as a result of nest trapping. In 1950, however, five females deserted their nests following trapping and banding. In all cases of nest desertion the hens were trapped six to eight days prior to the hatching date as indicated by an examination of embryos in eggs from the deserted nests. There seemed to be relatively little danger of desertion if birds were trapped as late as one to three days prior to the prospective date of hatching. Grouse were trapped on their nests by two persons approaching the nest on the exposed side and quickly placing a cotton-mesh netting entirely over the nesting site, including the brushy cover. The female was caught as she flushed into the netting. A net, 6 ft. by 12 ft. with 1-in. mesh, was used. This type of net was considered ideal for trapping purposes, and subjected the hens to a minimum risk of injury. No hens were injured during actual trapping, although one bird dislocated a wing while being handled. The females that were successfully trapped invariably sat tight until the netting was completely over the nest site; others, more wary, flushed wildly before the netting was in position.

Evaluation of the Trapping and Transplanting Program

Control of Damage

Sage grouse damage to crops has been reported on practically all agricultural areas thrust into the midst of extensive sage grouse habitat in the West. The heaviest damage to crops has apparently occurred in Wyoming, although other states have reported limited amounts of damage.

During years of high sage grouse abundance in the upper Green River Basin, claims totaling several thousand dollars have been submitted by ranchers in Eden Valley and in the Boulder-Pinedale region for damage inflicted to clover and alfalfa crops. During July, 1948, twenty-two damage claims, amounting to $4,500, were submitted to field representatives of the Game Commission by the farmers in the Eden Valley area. A special sage grouse hunting season was held in Eden Valley in 1948. This was followed by general open seasons in four southwestern Wyoming counties in 1950 and 1951. In addition to the reductions in numbers effected by the hunting seasons, Eden Valley grouse populations were further reduced by intensive live-trapping operations in 1949 and 1950. As a result of these management procedures a substantial number of the farmers withdrew their damage claims. Approximately $900.00 has been paid by the Game Commission on sage grouse damage claims during the past ten years.

In agricultural areas sage grouse feed extensively upon the leaves of alfalfa, clover, field peas, beans, and potatoes. Additional damage is inflicted by the birds upon hay crops as a result of trampling, which makes mowing difficult. While on an inspection tour of sage grouse habitat near Craig, Colorado, this writer talked with several landowners who complained of sage grouse damage to their gardens.

Very limited damage from sage grouse has been reported in Oregon. Control methods have consisted in placing the damage area within the boundaries of open hunting territory. Oregon has also resorted to live-trapping on a minor scale in order to remove birds from areas of high concentration. Sage grouse depredations in Idaho have been considered neither serious nor widespread. Some budding of field beans has occurred, and a little cropping has been done on alfalfa. Other than hunting seasons, the only control measures practiced have been a periodic flushing of the birds away from fields during the time that they could do damage. In Washington sage grouse damage has been confined to alfalfa and potato fields on irrigated lands within hunting areas. No damage complaints have been reported at any time in Montana. Conversely, most of the Montana farmers who have expressed opinions stated that they valued these birds for their grasshopper-consuming abilities.

Movements and Survival Following Transplanting

In 1950, prior to the sage grouse hunting season, approximately 1,700 birds were live-trapped and removed from ranches in Eden Valley. Releases were made near Dubois (325 birds), Sheridan (475 birds), and Midwest (400 birds). The remainder of the birds (approximately 500) were released at selected sites in territory which was to be opened for sage grouse hunting in the upper Green River Basin for purposes of making studies on movements and planting success. All birds were leg-banded prior to release. Hunters returned eighty-eight bands from birds killed during the 1950 open season.

Analyses of these band recoveries indicated that approximately 55 per cent of 152 adult males and 80 per cent of 52 adult females released at distances varying from 20 to 40 miles south of

TABLE 32
1950 *Recoveries of Sage Grouse Trapped in Eden Valley and Released to the Southward*

Date	Planting Site	Number and Sex of Released Birds				Number and Sex of Birds Recovered in Eden Valley			
		M*	J*	M†	J†	M*	J*	M†	J†
July 9, 1950	Gaston Crossing	24	0	2	0	2		1	
July 19, 1950	Green River	28	6	21	2	0		3	
July 20, 1950	14 Mile Hill	37	3	14	3	4		1	
July 22, 1950	15 Mile Spring	19	8	15	8	3		2	
		108	17	52	13	9		7	
Sept., 1947	14 Mile Hill			52				2	
Sept., 1949	Uinta County	132	88	143	151	1		1	
		132	88	195	151	1		3	

TABLE 33
1950 *Recoveries of Sage Grouse Trapped and Released in Eden Valley*

Trapping and Release Site	Releases				Recoveries			
	M*	J*	M†	J†	M*	J*	M†	J†
1949 Alfalfa fields	1	10	40	22	0	0	11	1
1949 On nests			14				3	
	1	10	54	22	0	0	14	1
1950 Alfalfa fields	20	29	96	35	2	7	17	8
1950 On nests			20				3	
1950 Strutting grounds	20				4			
	40	29	116	35	6	7	20	8

	1949 Releases	1950 Recoveries	1950 Releases	1950 Recoveries
Adult male	1	0 - 0%	40	6 - 15%
Juvenile male	10	0 - 0	29	7 - 24
Adult female	54	14 - 26	116	20 - 17
Juvenile female	22	1 - 5	35	7 - 20
Totals	87	15 - 17%	220	40 - 18%

*Male   †Female

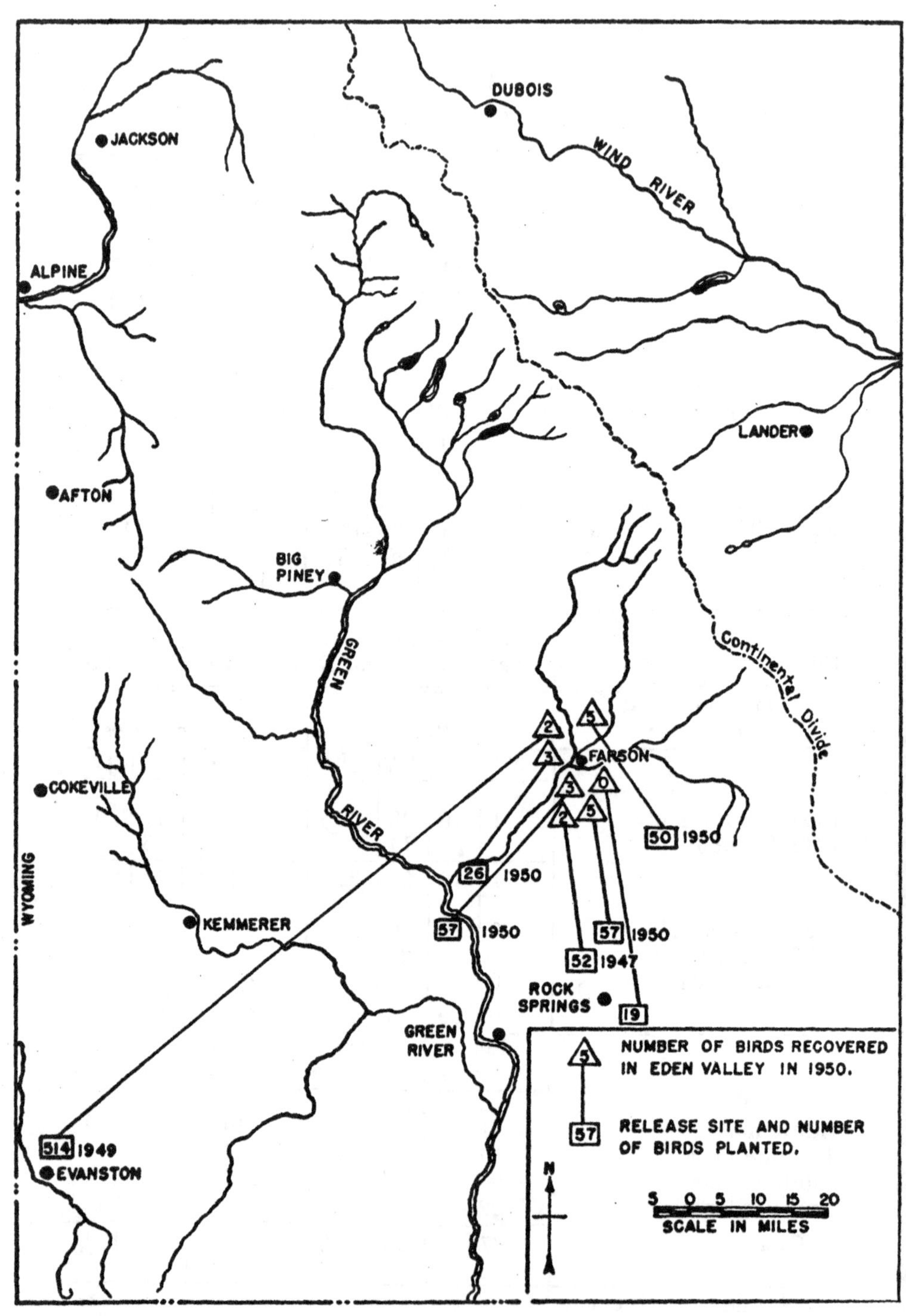

FIGURE 17. MOVEMENTS OF SAGE GROUSE RELEASED TO THE SOUTHWARD OF EDEN VALLEY

Eden Valley actually returned to the vicinity of their capture within the period from July 7 to August 26, 1950. A comparison of percentages of banded birds released to the southward and killed in Eden Valley and those released and killed in Eden Valley appears in Tables 32 and 33.

These band returns indicate that the sage grouse removed from Eden Valley and transplanted to other areas in the Green River Basin quickly dispersed from the site of planting. The majority of birds returned to the exact area from which they had been trapped. Others were killed at locations from 20 to 30 miles from the point of their release. Bands were recovered from sixteen birds which had returned to Eden Valley subsequent to releases varying from 20 to 40 miles from the site of their capture. These returns originated from five separate plantings involving 209 birds released at sites to the southward of Eden Valley (Figure 17). All band recoveries obtained in Eden Valley from these transplants involved mature birds which had returned within a maximum period of seven weeks after release. There was no indication from band recoveries that juvenile birds had returned to Eden Valley from any of the experimental release sites in the Basin.

It is considered significant that no direct recoveries were obtained locally from birds which had been transplanted into areas north of Eden Valley (Table 34). All birds known to have returned to Eden Valley during the same season as transplanted have originated from plantings made to the southward. The pattern of movement of adult sage grouse, after transplantation in the Green River Basin, further substantiates the belief that adult birds tend to migrate to the upper drainages during the summer and early fall period, especially since no direct recoveries (recovered same year as banded) were obtained locally from 739 birds released at eight separate sites to the northward of Eden Valley. Locations of these release sites are shown in Figure 18.

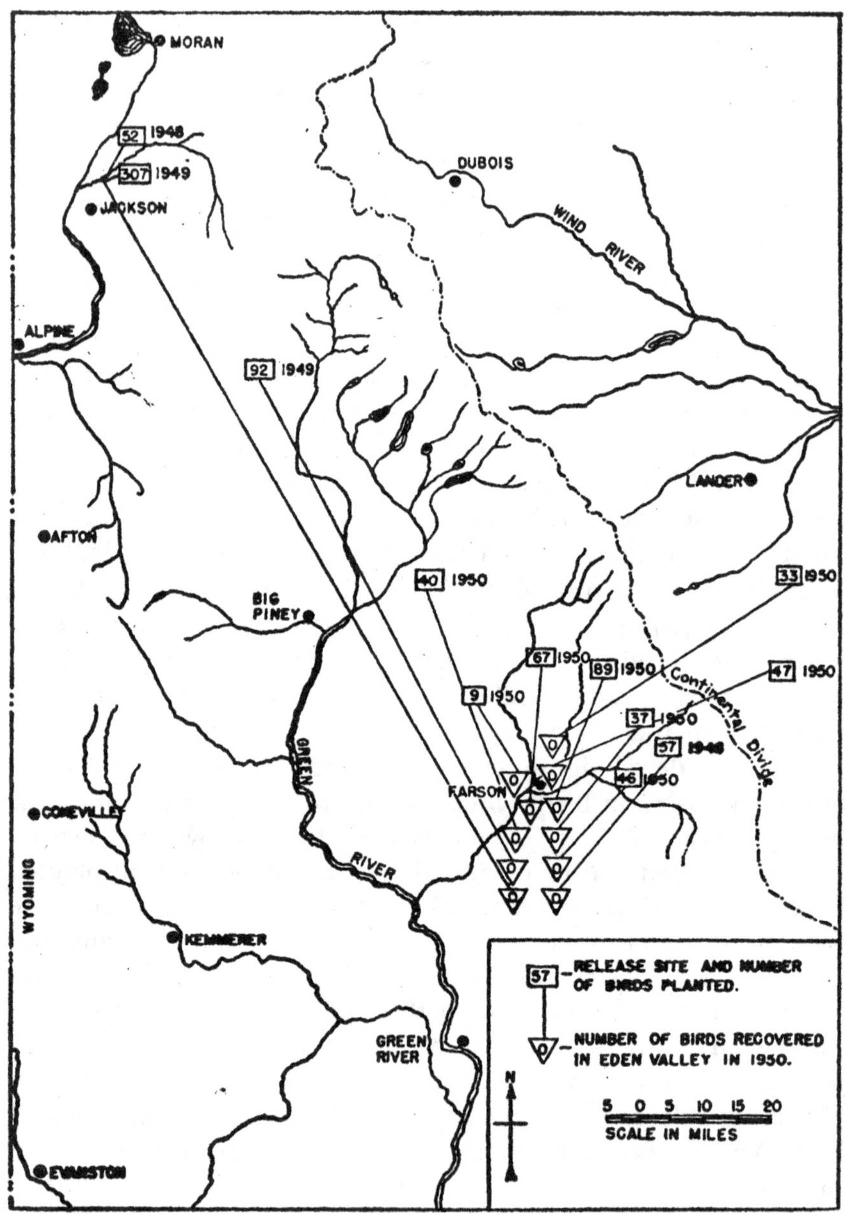

FIGURE 18.   MOVEMENTS OF SAGE GROUSE RELEASED TO
THE NORTHWARD OF EDEN VALLEY

TABLE 34
1950 *Recoveries of Sage Grouse Trapped in Eden Valley and Released to the Northward*

Date	Planting Site	Number and Sex of Released Birds				Number and Sex of Birds Recovered in Eden Valley
		M*	J*	M†	J†	
July 8, 1950	Sublette Springs	9				
July 14, 1950	Superior Oil Well	37				
July 23, 1950	Pacific Creek	2	22	5	17	No recoveries
July 30, 1950	Elkhorn	27	15	20	27	In Eden Valley
July 31, 1950	Sand Springs	15	2	21	2	
Aug. 1, 1950	Chalk Butte	23	16	6	22	
		113	55	52	68	
Aug., 1948	Teton County	40	2	8	2	No recoveries
Aug., 1949	Teton County	185	37	36	49	In Eden Valley
		225	39	44	51	
Oct. 1949	Hoback Rim	1	44	10	37	No recoveries In Eden Valley

*Male
†Female

Indirect recoveries (not recovered same year as banded) were obtained on an adult male and female which had returned to Eden Valley after their release near Evanston, Wyoming, in September, 1949, involving a return distance of more than 100 miles. Several bands were recovered from birds killed in Uinta County at distances from 20 to 30 miles from the point of release near Evanston. Two mature hens planted 26 miles south of Farson in September, 1947, were recovered in Eden Valley in September, 1950. All birds were originally trapped in Eden Valley, and were recovered during the hunting season from August 27 to September 10, 1950.

Juvenile male, No. 2666, was trapped in Eden Valley and released near the Hoback Rim on October 1, 1949, about 90 miles northwest of the point of capture. This same bird was shot by an Idaho hunter approximately 25 miles northeast of Soda

Springs, Idaho, on September 17, 1950. As nearly as can be determined, this individual bird must have moved an airline distance of not less than 75 miles and over at least two high mountain ranges from its point of release. In all probability, the movement followed drainage patterns, and was probably much greater than would be indicated by the airline distance.

During the spring of 1951 four indirect band recoveries were obtained from carcasses found in Eden Valley. These birds were banded in 1949 and 1950 and were presumed to represent crippling losses from the 1950 hunting season. One of these birds (a mature female) had returned to Eden Valley from Elkhorn, 30 miles north of the point of capture.

Twenty-two leg bands were recovered from birds killed in Eden Valley during the 1951 hunting season. These birds also had returned from 1950 release localities in the Green River Basin. Another adult hen was killed in the valley after being released near Evanston. The first band recovery from a bird moved to the east of the Green River Basin was recorded in 1951 when an adult hen returned to Eden Valley after release in the Wind River Basin near Lander, Wyoming. Several indirect recoveries were obtained locally in 1951 from birds released in 1950 to the northward of Eden Valley.

In 1948 and 1949 a total of 359 sage grouse was removed from Eden Valley and transplanted to the sagebrush flats in Jackson Hole, Wyoming. Based on strutting ground censuses, it was determined that approximately 500 sage grouse were resident in this area prior to the releases. In an attempt to evaluate the success of the Jackson Hole releases, several strutting grounds were closely observed in the spring of 1950 and 1951. The legs of birds on the strutting grounds were closely checked with binoculars for the presence of aluminum and colored plastic leg bands. These observations were made from both a car and a blind. Due to the lateness of the season, few females were present on the strutting grounds. In 1950, 11 banded cocks were observed on two strutting grounds on which 117 cocks were actually checked. The histories of these banded birds are recorded in Table 35.

TABLE 35

Histories of Banded Sage Grouse Observed on Jackson Hole
Strutting Grounds in 1950

Strutting Ground Location	Date	No Banded Birds Seen	Release History
Jackson Hole Airport	5/13	4 Adult Cocks	Released as juvenile cocks on Sept. 3, 1949, about two miles from the strutting ground.
Jackson Hole Airport	5/13	2 Adult Cocks	Released as juvenile or adult cocks Aug. 31, 1949, about 8 mi. north of strutting ground.
Antelope Flat	5/14	2 Adult Cocks	Released as adult cocks in Aug., 1948, about two miles south of the strutting ground.
Antelope Flat	5/14	2 Adult Cocks	Released as juvenile cocks on Sept. 3, 1949, about 8 miles south of strutting ground.
Antelope Flat	5/14	1 Adult Cock	Released as a juvenile or adult cock in Sept., 1949, near strutting ground.

In 1951, ten banded cocks and one banded hen were observed on four strutting grounds on which eight-nine cocks and ten hens were actually checked. Histories of these banded birds are recorded in Table 36.

Due to time limitations, several other strutting grounds in the Jackson Hole area were not checked either in 1950 or in 1951 for the presence of marked birds. At least six of the banded cocks observed in 1950 and three of the banded cocks observed in 1951 were known to have been released as juveniles.

Four of the banded cocks observed in 1950 and 1951 were originally released as adult birds. It was impossible to determine the age classes represented by the other banded birds since some of the adults and juveniles were marked similarly upon release. There were also indications on the legs of some birds that the

TABLE 36
Histories of Banded Sage Grouse Observed on Jackson Hole
Strutting Grounds in 1951

Strutting Ground Location	Date	No. Banded Birds Seen	Release History
Antelope Flat	5/25	1 Adult Female	Released in Jackson Hole in Aug., 1949.
Antelope Flat	5/25	1 Adult Cock	Released as juvenile male 8 miles south of the strutting ground on Sept. 3, 1949.
Jackson Hole Airport	5/25	2 Adult Cocks	Released in Jackson Hole in Sept., 1949.
One mile S.E. of Airport	5/26	2 Adult Cocks	Released as juveniles on Sept. 3, 1949, 3 miles east of strutting ground.
One mile S.E. of Airport	5/26	2 Adult Cocks	Released as adults on Elk Refuge, Aug. 26, 1949, about 4 miles east of strutting ground.
One mile S.E. of Airport	5/26	2 Adult Cocks	Released on Sept. 3, 1949, 3 miles east of strutting ground.
One mile S.E. of Airport	5/26	1 Adult Cock	Released 10 miles north of strutting ground on Aug. 31, 1949.

marking system had broken down, due to the plastic colored bands slipping over the aluminum bands.

From the standpoint of increasing a resident sage grouse population through planting of wild stock, it is of interest to note that in 1950, 9.5 per cent of the birds checked on the strutting grounds were banded, whereas in 1951, 11 per cent of the birds observed were banded. These observations indicated that relatively few of the transplanted birds had established themselves in resident flocks by the onset of the next breeding season. This conclusion was corroborated by the censuses of males on Jackson Hole strutting grounds from 1949 through 1951 (Table 20).

Sage grouse traps are constructed of 2-inch mesh, cotton netting erected on a rectangular frame 14 feet wide, 28 feet long, and 42 inches high. *Photo by Euvern Putnam*

The net structure is supported by three transverse arches set into angle-iron posts. *Photo by Putnam*

Posts are 52 inches long and pointed, with a crosspiece welded 10 inches above the base for support and measurement purposes. *Photo by Putnam*

Trap is secured to ground at 3-foot intervals with iron stakes fitted with a short piece of rod for holding the netting. *Photo by Putnam*

For strength and stability the netting is faced along the edges with ¼-inch cotton rope. Metal rings are fastened to the rope-facing at each of the four lower corners of the trap. *Photo by Putnam*

Each trap is equipped with two wings of 2-inch mesh, cotton netting 150 feet long and 18 inches high. Wings are supported by permanently attached wooden pegs spaced 6 feet apart. *Photo by Putnam*

Entrance panels are attached to the front corners of the trap and meet 6 feet inside the front arch to create a narrow V-type opening. A strip of lath supports an entrance about 10 inches wide at the ground level. *Photo by Putnam*

Grouse are transported to release sites in bird farm crates

The 1949 census produced 191 males. In 1950, 223 males were counted on the strutting grounds. These counts were taken after the release of 185 mature males and 37 juvenile males in the fall of 1949. The increase in male numbers in this instance amounted to 11.5 per cent. Based upon total numbers of males on all Jackson Hole strutting grounds in 1950 and 1951, about 47 per cent of the cock birds were checked for bands. If it is assumed that all of the marked cocks of questionable age were released as juveniles, the survival rate by the next spring's breeding season would have approximated 45 per cent for the 37 young birds released, and 2 per cent for the 185 mature birds released. Since birds known to have been released as adults were observed on the strutting grounds in both years, it was more likely that the juvenile and adult survival rates actually did not exhibit such wide disparity.

On the basis of band recoveries in Eden Valley and sight records in Jackson Hole, there was considerable evidence that immature birds were more inclined to acclimate themselves to new habitat and to remain in the general vicinity of the transplant site than was the case with adult birds. The high percentage of recoveries on adult birds which had returned to Eden Valley and the extremely low percentage of sight records on mature birds released in Jackson Hole indicated that the mature age classes were highly intolerant to establishing themselves in the locality of a transplant site, regardless of its apparent suitability. Allred (1950) has reported that similar behavior patterns were displayed by immature and adult age classes of elk following their transplant to new areas in Wyoming.

In view of the pronounced seasonal migrations exhibited by sage grouse, we may expect a low degree of success to be associated with transplanting this species southward during the spring and summer seasons, and possibly at other times as well. Bands recovered during future hunting seasons may contribute additional information on the movements of the birds transplanted to the northward of Eden Valley. In any event there appeared to be strong tendencies for adult sage grouse of both sexes to disperse widely from areas in which they were liberated. It would have been desirable to have known the breeding status of the mature hens which were transplanted, since a significantly larger percentage of recoveries were obtained from adult hens (13.5 per cent) than from adult cocks (8.3 per cent). The urge to rejoin their broods may have accounted for the higher percentage of recoveries of mature hens. Since no sage grouse hunting seasons have been held since 1936 in areas of Wyoming lying to the eastward of the Green River Basin, it has been practically impossible to obtain recoveries from birds transplanted into other regions of the state. Prior to 1951, there had been no Eden Valley recoveries from birds transplanted onto major watersheds to the eastward of the Green River drainage.

In addition to the birds which were banded and transplanted into areas outside of Eden Valley, over three hundred birds were banded and released at the sites of their capture in Eden Valley in 1949 and 1950. By comparison of the ratios of banded and unbanded birds in the total kill (use of Lincoln's Index) a fairly reliable estimate of prehunting season sage grouse numbers in Eden Valley was obtained. The birds which were banded and released were obtained from three sources, namely: live-trapping in alfalfa fields, live-trapping on strutting grounds, and nest trapping of females (Table 33).

Utilizing the 1950 data on releases and recoveries in Eden Valley, the prehunting season population of sage grouse in Eden Valley was computed by the Lincoln Index at 33,000 birds. A spring breeding population of about 7,500 females and 2,500 males was in Eden Valley in 1950. Studies of nesting success and brood survival indicated that 54 per cent of the nests hatched successfully and by mid-August these broods contained about 4 young. It can, therefore, be concluded that approximately 16,000 young birds would have been present in Eden Valley at the onset of the 1950 hunting season. According to the banding returns there were about 14,600 juvenile birds in Eden Valley at the beginning of the hunt. Prior to the hunting season 1,226 adult birds and 534 juvenile birds were removed from Eden Valley by transplanting operations. The theoretical adult : juvenile ratio at the time of hunting would have approximated 100 adults to 159 juveniles. The actual ratio encountered at the game checking stations was 100 adults to 133 juveniles. The actual adult : juvenile ratio indicated that considerably more adult birds were in Eden Valley during the hunting season than had been there in the spring. It seemed that approximately 1,500 birds were involved in a movement into agricultural areas. This sort of behavior could be expected in view of the abundance of alfalfa and water which was available during the dry periods of midsummer, and the fact that the cultivated areas were located in the midst of practically unlimited sage grouse habitat. It should be emphasized again that the boundaries of the Eden Valley area have been arbitrarily established at the limits of the irrigation district. Studies of sage grouse flock movements and populations have been conducted on the basis of this area as defined.

There was considerable variation between the total population in Eden Valley as obtained by the Lincoln formula and that obtained from field studies and age ratios. The actual number of sage grouse in Eden Valley at the beginning of the hunting season was believed to be in the vicinity of 25,000 birds. The accuracy of the population figure derived from Lincoln's formula was contingent upon all of the 220 banded birds surviving, or at least remaining in the area, until the hunting season. This was unlikely since the birds were banded throughout the period from May 9 to August 26, 1950. Recoveries of bands from these 220 birds accounted for 18 per cent of the birds banded, whereas 24 per cent of the 62 birds which were banded and released during August 1950 were recovered during the hunting season. This figure was considered to more accurately represent the percentage kill, and, if so, would have meant that the actual population was about 25,700 birds.

It is noteworthy that 8.3 per cent of the 108 adult males transplanted to the southward of Eden Valley were recovered near the original trapping site, as compared with 15 per cent recoveries on adult males released in Eden Valley. There were 13.5 per cent local recoveries on 52 adult females which were transplanted to the southward, whereas 17 per cent of the adult females released locally were recovered. This would have indicated that about two-thirds of the adult birds transplanted to the southward of Eden Valley returned to the vicinity of their capture within a two-month period.

Conclusions

A compilation of trapping and transplanting data for the 1949 and 1950 seasons appears in the Appendix. Most of the birds were trapped from alfalfa fields on small ranches in Eden Valley. All of these birds were released in Wyoming.

Trapping and transplanting of sage grouse in the Eden-Farson agricultural area of northern Sweetwater County was more successful in 1950 than in any previous year of trapping. During the period from July 7 to August 26 and from October 4 to October 26, 1950, a total of 2,053 sage grouse were trapped. The greatest number previously taken in any one season was 1,123 in 1949. An average of 63 grouse were captured on each of twenty-seven trapping days prior to the hunting season, in contrast to an average of 29 birds per trapping day following the hunting season. Maximum daily catches were 129 and 137 birds. Catches of more than one hundred birds were made on six separate days. Eight hundred birds were removed from three alfalfa fields spaced within a distance of slightly more than 1 mile. Four hundred grouse were taken from a 40-acre alfalfa field on the McMurray ranch. Nearly seven hundred birds were trapped on the McComas ranch in the Eden area. Several ranches in the Eden area were plagued with hundreds of birds throughout the summer of 1950. Trapping operations were concentrated on these ranches. Damage to alfalfa fields in the Farson area was negligible in comparison to the Eden district. This was the result of smaller numbers of birds and the large parcels of land which had reverted to alkali in the Farson segment. Numbers of birds captured and respective trapping sites in Eden Valley are shown in Figure 3.

In spite of mass trapping operations in 1949 and 1950 and an extensive hunting season in Eden Valley in 1950, sage grouse flocks, particularly mature males, continued to damage Eden Valley alfalfa fields in 1951. Several fields of 40 acres or less were frequented daily by from five hundred to a thousand birds. As long as sage grouse populations remain at peak levels in the upper Green River Basin, there must be effective and practicable methods of reducing the damage to alfalfa fields. Under the present day concepts of management (short hunting seasons and low bag limits) and considering the low human populations characteristic of most areas in Wyoming, it is doubtful if sufficient hunting pressures can be exerted on sage grouse populations in the Basin to entirely eliminate the damage potential to alfalfa fields in isolated agricultural areas. A supplementary method, such as intensive live-trapping on a selective basis, should be utilized to effect local control. In this way large numbers of birds which are actually doing damage can be removed from localized damage areas in the summer, at a period when hunting seasons are neither desirable nor practicable.

It was apparent from the 1950 trapping season that live-trapping of sage grouse is an effective method of reducing the number of birds and damage potential in localized problem areas, *provided* the birds are transplanted onto entirely different river drainages. It was also evident that planting stock easily can be obtained in large quantities through the trapping of wild populations. We can expect relatively little success to be associated with attempts to establish transplanted birds (especially adults) in a specific location.

In summary, it may be stated that more than thirteen thousand sage grouse were removed from the Eden Valley irrigation district from 1949 to 1951 by means of live-trapping operations (1949 and 1950) and authorized hunting seasons (1950 and 1951). On the basis of total acreage in the Eden Valley irrigation district, an average annual harvest of twenty-one birds per square mile was taken each year from 1949 through 1951.

RECENT SAGE GROUSE HUNTING SEASONS
IN WYOMING

Nature of the Seasons

On August 27 and 28, 1950, a general open season on sage chickens was conducted in Sweetwater, Uinta, Lincoln, Sublette, and Fremont counties of southwestern Wyoming, with a four-day extension (covering the week-end days of September 2 and 3, and September 9 and 10) in the immediate area of the Eden Valley irrigation district in northern Sweetwater County; the latter provision was made in an effort to further reduce sage grouse populations damaging alfalfa crops in that region. With exception of a controlled season on mature males in Eden Valley from August 1 to 10, 1948, this marked the first year in which a general open season on sage chickens had been held in Wyoming since 1936. The 1951 hunting season was held on August 18 and 19 and embraced the same territory as the 1950 season.

242

Determination of the Kill

During both years game checking stations were established on the main highway and on secondary roads for the purpose of determining the sage chicken kill and hunting pressure on areas included within and adjacent to the sage grouse study tracts. In 1950, approximately 140 square miles were opened to hunting within the Eden Valley district. Included in this area were 9,000 acres (14 square miles) of cultivated land which were located in the center of extensive sage grouse habitat. Hunters were required to check in and out of the Eden Valley hunting area in 1950. During the six days of the 1950 season, 6,179 birds were killed in the Eden Valley district to give a kill of approximately 45 birds per square mile of hunting territory. This did not include crippling losses and illegal kill (old birds abandoned in the field). The number of birds bagged was considered to represent a 25 per cent harvest of the total prehunting season population, and was based upon returns from banded birds, strutting ground and nesting censuses, as well as a knowledge of local hatching success and brood survival. At the onset of hunting, flocks of mature cocks and the hens without broods immediately withdrew from the agricultural areas and took refuge in outlying sagebrush tracts. The majority of brood hens and young birds were more inclined to take refuge in the hay and grain fields and, thus, were exposed to a constant hunting pressure throughout the season. This pattern of movement was reflected in the high percentage of mature hens and young birds which appear in the total kill for both years (Table 37).

The two-day 1951 season produced a kill of slightly more than 4,000 birds in Eden Valley, as compared with a kill of 2,200 birds on the opening two days of 1950. Hunting pressure in Eden Valley increased from 872 hunters for the first two days of 1950 to 1,800 hunters during the two-day season of 1951. It should be emphasized that the opening two days of the 1950 season were on Sunday and Monday, whereas the 1951 season was confined in its entirety to Saturday and Sunday. All hunting pressure in the later year was necessarily concentrated over a full week end. Considerably more publicity was also afforded the 1951 season than was the case in 1950.

On August 27 and 28, 1950, about 1,200 birds were estimated to have been killed in all of the outlying areas adjacent to Eden Valley in northern Sweetwater, southeastern Sublette, and southwestern Fremont counties. The 1951 kill in this same area was estimated at 1,600 birds from check station records. Some 4,000 square miles of sage grouse habitat are included in this vast area. The total kill for this extensive tract hardly exceeded the number of birds which existed on perhaps one-half township in almost any area selected between Eden Valley and the Wind River Range. These figures do not include birds killed in the Boulder-Pinedale area. Nesting studies had actually revealed larger numbers of sage grouse in the mountain foothills country than in Eden Valley. However, in all regions except Eden Valley, the annual kill in either year was noticeably small as compared to the number of birds that were available for harvest. Several thousand square miles of additional hunting territory in Lincoln, Sublette, and Uinta counties would likewise have fallen in this category.

Ageing and Sexing

Age and sex data were obtained on 2,599 birds representing 35 per cent of the birds checked through game stations in 1950, and on 2,058 birds representing 45 per cent of the birds checked in 1951 (Table 37).

An examination of the birds at game checking stations in 1950 disclosed an age ratio of 100 adults to 133 young. The adult : young ratio in 1951 increased to 100 adults : 142 young, indicating a more successful hatch, an increased brood survival, or both. These age ratios were considered to be closely representative of sage grouse populations in Eden Valley. The postbreeding movements of small numbers of adult birds into Eden Valley agricultural areas operated to lower the ratio of young birds to adults. Sage grouse young of the year were not present in sufficient numbers to constitute the normally higher proportion of the harvest as encountered in other species of upland game birds which exhibit a higher breeding potential.

Age determination of birds during the hunting season was accomplished primarily by an examination of the sternum and by testing the breaking strength of the lower mandible. The sternum of mature birds is completely ossified throughout its length and does not bend under moderate pressure exerted by the index finger. The tip of the young bird's sternum is extremely cartilaginous and will bend readily when pressed.

TABLE 37

Age and Sex Composition of the Sage Grouse Hunting Season Kill in Eden Valley—1950-1951

	1950		1951	
	Number of Birds	Percent of Kill	Number of Birds	Percent of Kill
Adult males	351	14%	231	11%
Young males	683	26	537	26
Adult females	763	29	619	30
Young females	802	31	671	33
	2,599	100%	2,058	100%

There is also a distinct difference between adult and young birds in the texture and character of the plumage. The feathers of immature birds are considerably softer and less compact. Young birds of both sexes exhibit an irregularly defined patch of finely spotted juvenile feathers on the neck and breast region, which is distinct from adjacent feathers and presents a pattern of contrasting plumage on the breast. Adult hens possess a solid feather pattern across the breast with no evidence of spotted feathers. Adult males are easily distinguished by the stiff white feathers on the breast and by their large size. The toes of young birds are greenish-yellow in color in contrast to the greenish-black toes of mature birds.

Sex determination of young birds during the hunting season is accomplished mainly by visual inspection of the gonads, size and shape of the head, and length of the middle toe. Sage grouse invariably are field-dressed prior to arriving at game checking stations, hence the body cavity is open for visual examination. The testes of the male are bluish-black oval bodies lying on either side of the backbone just anterior to the kidney region. In the female, the ovary, resembling a cluster of minute fish eggs, lies in the same position as the testes of the male. In the sage grouse, as with most birds, only the left ovary develops. On young birds of the same age the middle toes of males will average a full toenail-length longer than the middle toes of the females. The difference in size and shape of the head is a relative quantity, dependent likewise upon the age of the young birds. As the birds grow older, the heads of young males will appear bulkier and longer than young females. It should be emphasized that age and sex determination of sage grouse based on plumage characters and on the size of the head and the middle toe would have not been reliable without the experience gained from handling several thousand birds during live-trapping and banding operations. The sternum and mandible tests for ageing and examination of gonads for sexing would seem to offer the most reliable techniques for game personnel not accustomed to handling a large series of sage grouse.

Hunter Success

The daily bag limit in both 1950 and 1951 was three birds. Individual hunter success during both years was high; practically all hunters who expended a little effort were rewarded with full bag limits. According to hunter reports, an average of slightly less than one hour was required to kill a bird during the 1950 season in Eden Valley. The average kill per hunter on opening day in 1950 was 2.60 birds for all areas combined. In 1951 it dropped to 2.28 birds per hunter, due mainly to lower hunter success in the Eden Valley area. It was rather significant that the average kill per hunter in Eden Valley dropped from 2.7 birds on opening day in 1950 to 2.2 birds on opening day in 1951, reflecting a probable decrease in total production as a result of last year's early season and the accompanying heavy kill of female birds. Although the average kill per hunter on opening day in Eden Valley decreased in 1951, the average opening day bag in the outlying areas increased from 2.4 birds in 1950 to 2.53 in 1951. Females aggregated 60 per cent of the 1950 kill. Hens composed 63 per cent of the 1951 kill *primarily* as a result of opening the season ten days earlier. The average daily kill per hunter for the Eden Valley area is presented in Table 38.

TABLE 38
Average Daily Kill of Sage Grouse Per Hunter in Eden Valley—1950-1951

Date of Season	Aver. Daily Bag
Aug. 27, 1950	2.70 birds
Aug. 28, 1950	2.30
Sept. 2, 1950	2.13
Sept. 3, 1950	2.05
Sept. 9, 1950	2.29
Sept. 10, 1950	1.95
Aug. 18, 1951	2.22
Aug. 19, 1951	2.00

During the 1950 six-day hunting season in Eden Valley 6,179 sage chickens were killed. In two days of hunting on the same area in 1951 approximately 4,000 birds were killed.

Individual hunter success was high, and the majority of hunters commented on the large numbers of birds encountered in the field. Little effort was required by hunters to fill their bag limits. A summary of the 1950 sage chicken season in Eden Valley appears in Table 39.

Hunting Pressure Trends

Within the past few years there has developed considerable opposition on the part of both game departments and sportsmen to the opening of fairly large areas of land to sage chicken hunting. This reluctance has been attributed to a fear that annual hunting pressures would be too heavy to maintain sage grouse populations at desired levels of abundance. These factions have believed that sage grouse populations should be protected for additional years in the belief that their numbers would continue to increase. There is a widespread opinion among sportsmen that game and fish populations in an area belong to the local populace, and that by opening these areas to hunting and fishing, a host of out-of-county sportsmen will move in and usurp the sporting and aesthetic rights they have ordained to be their own. As applied to sage grouse in Wyoming, these ideas have little merit, either on the basis of fact or law.

TABLE 39

Summary of the 1950 Sage Grouse Hunting Season in Eden Valley
(Bag and Possession Limit, Three Birds)

BASIC DATA	Acreage hunted	89,600
	Number of hunters	2,566
	Number of gun-hours	5,543
	Kill: Sage grouse	6,179
HUNTING PRESSURE	Hunters per 100 acres	2.8
	Gun-hours per 100 acres	6.4
	Acres per hunter	35.0
	Acres per gun-hour	15.6
KILL PER UNIT AREA	Sage grouse per 100 acres	7.0
	Acres per sage grouse killed	14.5
KILL PER 100 HUNTERS	Sage grouse per 100 hunters	240.0
	Sage grouse per hunter	2.4
HUNTERS PER UNIT KILL	Hunters per sage grouse killed	.42
KILL PER 100 GUN-HOURS	Sage grouse per 100 gun-hours	111.0
GUN-HOURS PER UNIT KILL	Gun-hours per sage grouse killed	.88

In support of the foregoing conclusion, the following discussion is presented from an analysis of data collected at game checking stations in 1950 and 1951.

On August 27 and 28, 1950, 92 per cent of the hunting pressure in the Eden Valley area was of local origin, i.e., originating in Sweetwater County. This percentage is based on the locality of car registration and not actual residence, so it is undoubtedly a conservative figure. Only 4 per cent of the total hunting pressure originated in the Lander-Riverton region which is the population center nearest to the territory opened to sage grouse hunting. Approximately two thousand hunter-days were expended in the areas previously described during the opening two days of the 1950 season (Sunday and Monday). Approximately 2,600 hunter-days were expended during the 1950 six-day hunting season in Eden Valley. Only 1 per cent of the hunting pressure originated with out-of-state hunters, and 87 per cent of the nonresident hunting was done on the latter two week ends coincident with the antelope hunting season. With the exception of hunting by nonresidents, all of the hunting pressure during the last two week ends in Eden Valley was essentially of local origin. The three-bird possession limit, combined with the long distances involved in travelling to hunting areas, was responsible for the lack of interest shown by out-of-county hunters. The nearest population centers are from 40 to 75 miles removed from Eden Valley. From license-sale trends in recent years, it is estimated that at least 3,500 Sweetwater County residents were in possession of licenses entitling them to hunt sage grouse. The total population of the county as listed in the 1950 census report was 22,000 people. On opening day in 1951, 1,081 hunters were actually checked in the field. Making a liberal allowance for unchecked hunters in isolated areas of Sweetwater County, it is evident that only about 10 per cent of the people in the county, and substantially less than one-half of the potential bird hunters in the county, were in the field on the first day of the season. The hunting pressure decreased by 50 per cent on the second day of the 1950 season (Monday) and remained essentially at that reduced level over the next week end (Saturday and Sunday). Hunting pressure on the last week end dropped to about 25 per cent of that encountered on opening day.

Except for an increased participation by local hunters, the pattern of hunting pressure, as well as of areas opened to hunting, remained unchanged in 1951. The opening two days fell on Saturday and Sunday, and no additional hunting days were authorized in 1951. Consequently, hunting pressures on opening day increased about 10 per cent over that recorded in 1950. Increased hunter-participation was further caused by more effective publicity afforded the hunt in 1951 and the good hunter success which accompanied the 1950 season. As sage chicken seasons are continued in future years, it may be expected that the sport will become more popular. Many of Wyoming's pheasant areas are so far removed from hunters living in the better sage grouse areas as to make them relatively inaccessible. This situation promises to turn many prospective pheasant shooters to sage grouse. The declining trend of hunting pressure exerted on successive week ends indicates that the average sage grouse hunter becomes tired of the sport in a relatively short time.

The total human population in the four counties opened to sage chicken hunting in southwestern Wyoming is 40,656 persons. Exclusive of national forest lands, almost 12 million acres of sagebrush and salt-desert shrub types are located in the counties of Sublette, Lincoln, Uinta, and Sweetwater. In autumn, sage chickens are distributed over practically all of the non-timbered areas. Assuming that 10 per cent of the total population in this region was to take advantage of sage grouse hunting privileges, there would be only one hunter for every 4½ square miles of potential hunting territory. This unbelievably light hunter-density would decrease to still lower levels, once the thrill of opening day had been satisfied.

By their very nature the majority of land areas supporting moderate to heavy concentrations of sage grouse are located in the regions, and in most cases in the states, with the lowest per capita : land area ratios. In such areas hunting pressures are so light that it is difficult to obtain an adequate harvest of sage chickens. In those states situated on the periphery of the birds' natural range, hunting seasons must be closely controlled on a special permit basis due to the restricted areas available for hunting, to the small and localized populations of sage grouse, and to their proximity to large urban areas. In recent years California, Oregon, Washington, and Utah have all held successful seasons of this type (Table 2).

As for the belief that sage grouse should be completely protected for several more years in the hope that their numbers will continue to increase, it may be stated that in most areas of Wyoming the species was extended complete protection from 1936 to 1948. Records indicate that during the early years of World War II the birds were present in numbers sufficient to justify opening the season in Wyoming. The birds evidently declined in numbers during the mid-1940's, and then recovered by 1951 to reach what was probably their greatest peak of abundance since the early 1930's. Similar trends have been reported in other states. On most areas of its range, years of protection and of increasingly good management of western range resources have restored sage grouse to just about the highest levels possible under the present land-use policies.

It is important to realize that once game animals have increased to levels of abundance permitting a harvest, a policy of continued protection and further stockpiling does not necessarily result in increased yields and higher numbers of animals. In view of the precariously low levels formerly attained by sage grouse throughout its range and of the widespread belief that the species was on the verge of extinction, there is little wonder that game management agencies have been hesitant to relax the stringent protective measures which assisted in preserving the bird from extermination. It is the same philosophy as fostered by the buck deer law, by the belief that game refuges are inviolate, and by continued use and expansion of game bird farms and fish hatcheries. Game departments go all out in selling the sporting public a management idea, and then they experience extreme difficulty in modifying that idea to meet changing conditions.

Relation of Growth and Development
to Hunting Season Dates

It has been traditional in the West to commence hunting sage chickens during the summer. A check of the 1901 reports of the Wyoming Game and Fish Commission revealed that the first legally established sage chicken hunting seasons in the state were held from July 15 to October 15. Other states originally set their seasons along similar patterns. For example, Nevada reported that their first season on sage chickens was held from July 1, 1901 to March 1, 1902. The pattern of Nevada's sage grouse hunting seasons since 1901 is considered so typical of conditions throughout most of the birds' range that a summary of Nevada's seasons and bag limits are included in this report (Table 40).

TABLE 40

Summary of Sage Grouse Hunting Seasons in Nevada—1901-1951

Year	Season	Bag Limit	Year	Season	Bag Limit
1901	July 1—Mar. 1, 1902	?	1935	Aug. 11-12	?
1909	July 15—Oct. 1	10	1940	Aug. 4- 5	5
1915	July 15—Feb. 15, 1916	10	1941	Aug. ?	5
1917	July 15—Sept. 1	10	1942	Aug. ?	5
1920	Aug. 1—Aug. 31	?	\multicolumn{2}{No seasons held from 1943-1948}	
1923	Aug. 16—Aug. 31	10	1949	Aug. 21-22	5
1925	Aug. 1—Aug. 16	10	1950	Aug. 13	5
1931	Aug. 7—Aug. 9	10	1951	Aug. 19	5
1934	Aug. 19	?			

Until recent years sage chicken seasons in most states have
been set primarily for the benefit of the hunters. The welfare
of the bird itself has been given little consideration. Certain individuals, influential in the management of wildlife resources,
contend that hunters will habitually discard mature birds once
they are killed. They consider both adult and young birds to be
unpalatable and unfit for eating after mid-August, and maintain,
therefore, that sportsmen will not tolerate sage chicken seasons
later than early August. These minority groups continue to
exert pressure for early seasons so that the hunters may be able
to distinguish young birds from adults in the field. This type of
hunting produces a highly selective kill which is undesirable.

The issue of palatability arises from the fact that as the young
birds approach maturity their diet shifts more and more from
insects and green forbs to sagebrush leaves. There is little question but that the younger the offspring of a domestic or wild
animal the more tender will be the meat. However, the flavor
and palatability of meat generally improve as the animal grows
older.

There is a quantity of evidence from many wildlife studies to
demonstrate that early shooting seasons on game birds such as
sage grouse and waterfowl, both of which have breeding grounds
of a specialized type, do irreparable damage to the brood stock

A juvenile hen and an adult hen in October

A juvenile hen and a juvenile cock in October

A mature hen in October

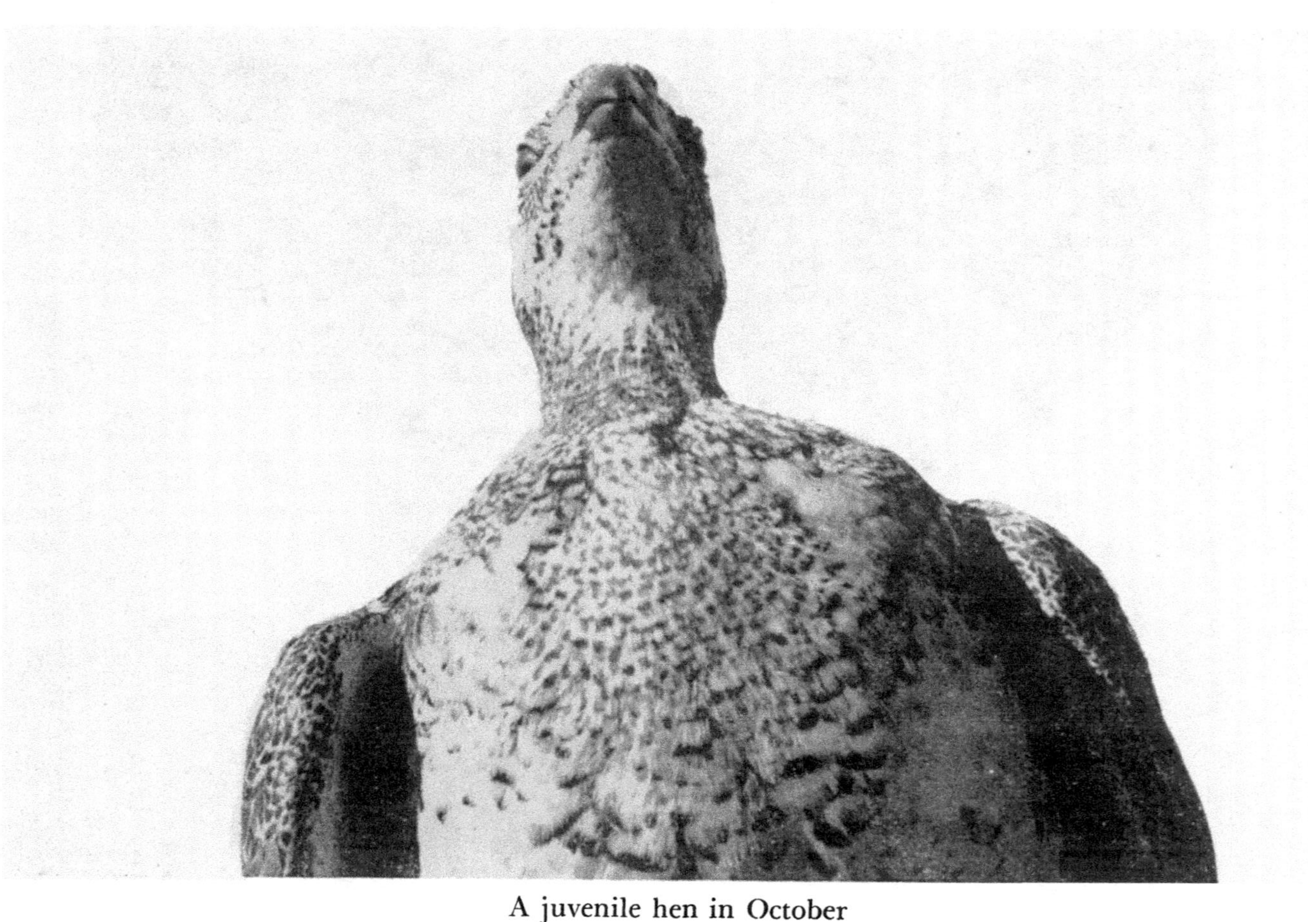

A juvenile hen in October

A juvenile cock in October

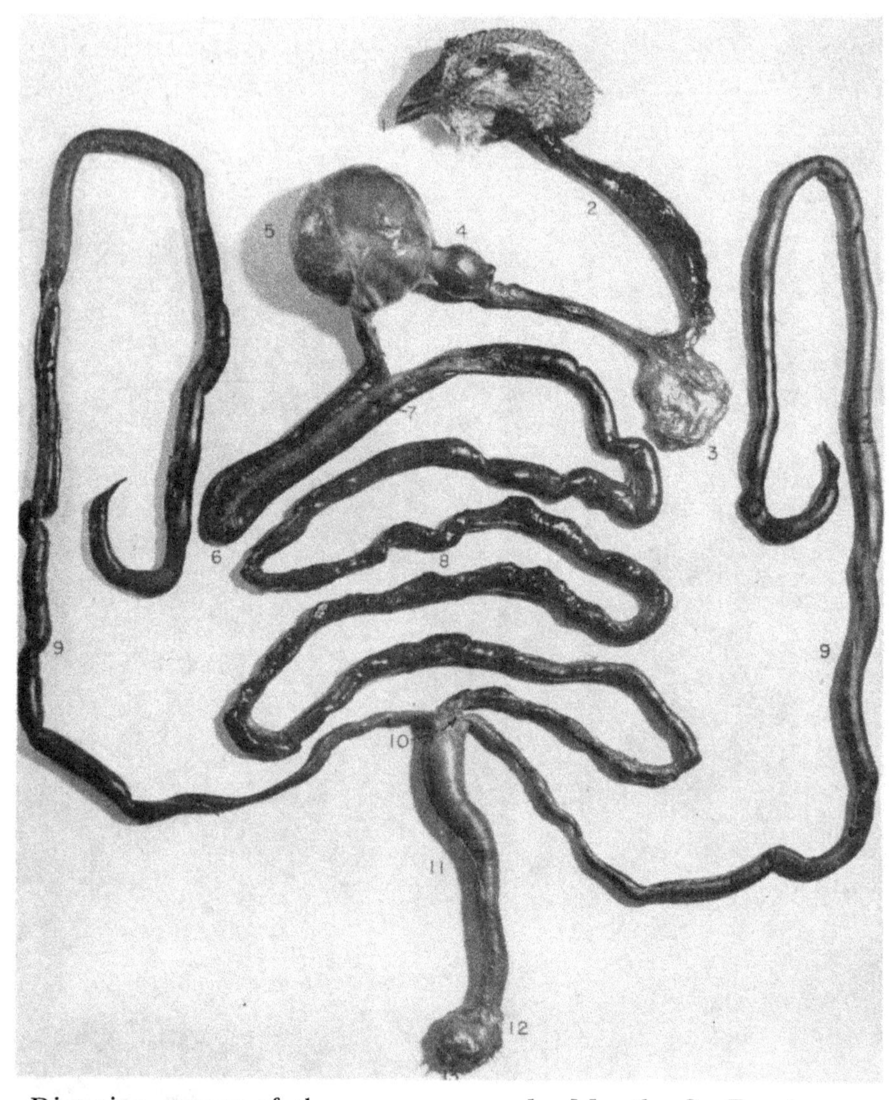

Digestive system of the sage grouse. 1—Mouth; 2—Esophagus; 3—Crop; 4—Proventriculus; 5—Gizzard; 6—Duodenal loop; 7—Pancreas; 8—Small intestine; 9—Caeca; 10—Caecal junction; 11—Large intestine; 12—Cloaca; 13—Vent.

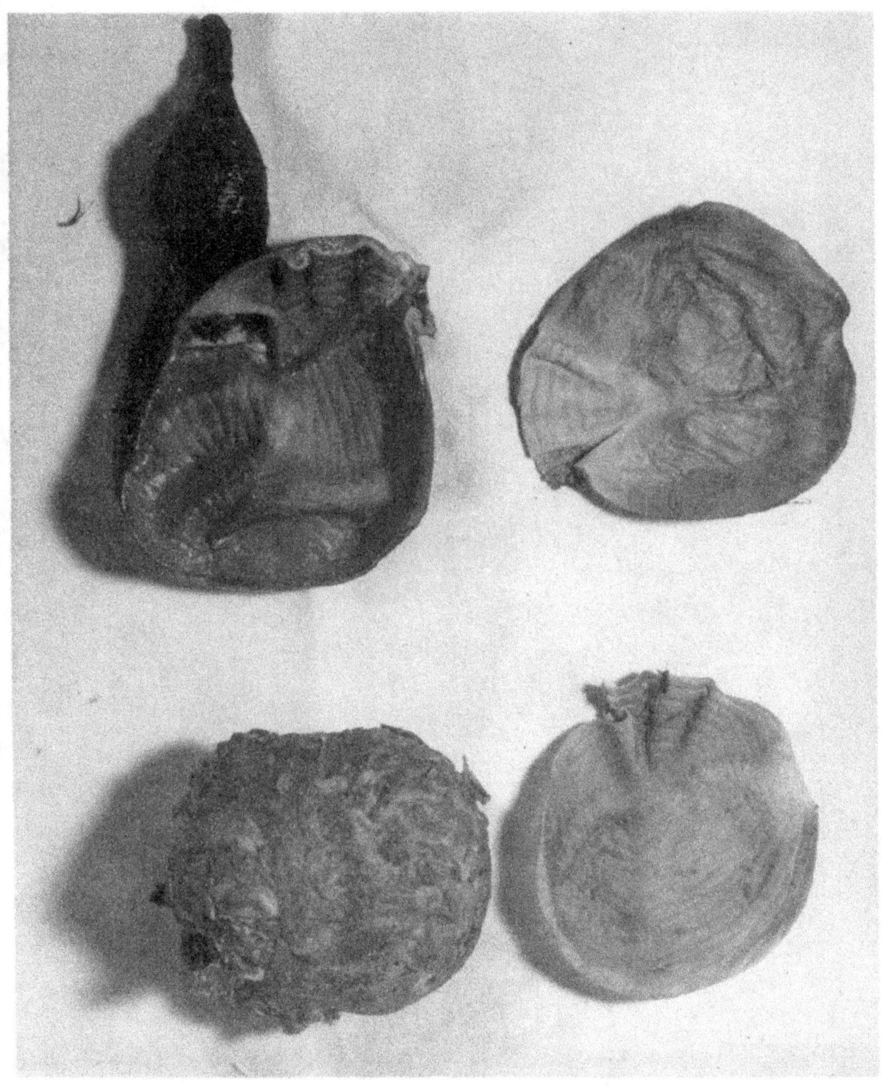

Cross section of a sage grouse gizzard showing thin-walled muscular tissue and horny inner lining—characteristics of the true gizzard. Food mass of sagebrush leaves

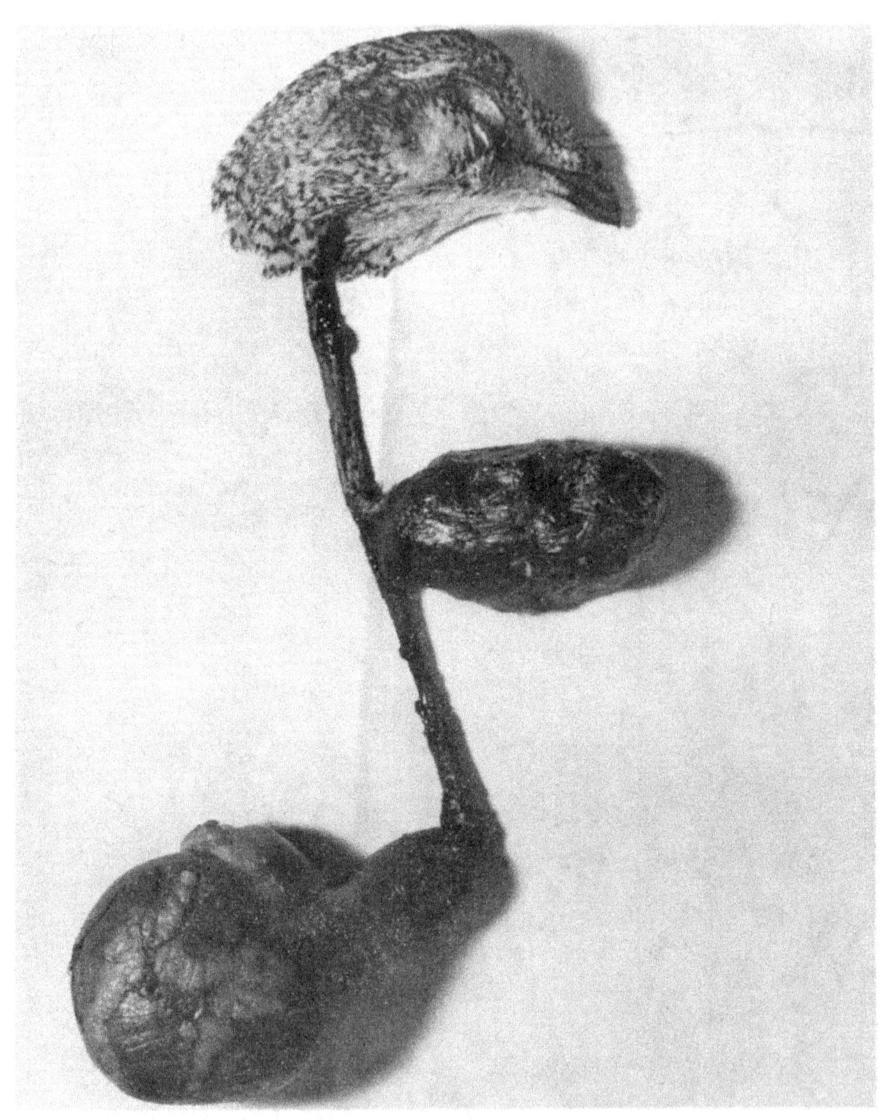

Crop and gizzard of a sage grouse

and young birds. In the case of these species, hunting pressure is concentrated on their breeding grounds at a time when the birds are in the moult, incompletely developed, and heavily concentrated on relatively small areas of land. Intensive early season hunting on the actual breeding grounds, before the hens and juvenile birds have dispersed onto their more widely distributed fall ranges, can "burn out" a nesting area, eliminating practically all of the birds so that few ever return. An important phase of sage grouse management is to assure that the birds mature on the site where they are hatched and commence their fall dispersal without a heavy kill-off, thus assuring the return of breeding birds to their home grounds in the spring.

The results of the 1950 and 1951 hunting seasons in Eden Valley have demonstrated the validity of these theories. With an August 27 opening in 1950, adult and juvenile hens composed 60 per cent of the total kill. The average kill per hunter was 2.7 birds on opening day. In 1951 the season was opened 10 days earlier (August 18), and adult and young hens made up 63 per cent of the total kill. The average kill per hunter declined 18 per cent below the 1950 level. It is believed that the nesting densities had dropped to correspondingly lower levels in view of the high percentage of females killed in 1950 and the reduced average kill per hunter in 1951.

A sex ratio of 44 males to 56 females was obtained on a series of 517 birds examined by Utah game department personnel during the hunting season conducted in that state on September 15 and 16, 1951. The effects of later hunting seasons in reducing the ratio of hens killed are also evident in these figures.

By now it should be an old and well-established principle that the *primary* consideration in game management should always be the welfare of the game species, with the sportsmen receiving an important but secondary consideration. Issues of palatability, law enforcement, damage control, etc., as related to time of game harvest, must necessarily be resolved without sacrificing the primary objective of management.

It is becoming increasingly evident that a long series of extended early shooting seasons results in a proportionately higher kill of hens than occurs under a system of hunting sage grouse in the fall. The traditional early hunting season has persisted, notably in Wyoming and Nevada. The net effect upon sage grouse populations over a period of time may, to our sorrow, bear surprisingly close resemblance to that encountered in former years. There is little justification for continuance of a policy that is now known to be so injurious to the continued survival of the species. It is the firm belief of many qualified biologists and game managers that annual sage grouse hunting seasons will not be possible in Wyoming or any other state under a continuing policy of early season shooting.

It has been especially gratifying to note that in recent years several western states have modified their policies to provide for substantially later hunting seasons on sage grouse. Within the past year the states of California, Idaho, Oregon, Utah, and Washington have authorized either September or October seasons for sage grouse, indicating the recent trend in the management of this valuable game resource.

During the 1950 and 1951 sage grouse hunting seasons in the Green River Basin, hunters commented on the small size of young birds which they killed in the regions lying at higher elevations near the foothills of the mountains. On the basis of hatching dates, these juveniles would have averaged from two to three weeks younger than birds in Eden Valley. It will be noted from the growth and weight curves of Eden Valley birds presented in Figure 10 that in mid-August juvenile males had attained 48 per cent of the weight of adult males and were less than half grown. At this same period juvenile females had attained 58 per cent of the weight of adult hens and were less than two-thirds grown. Adult males averaged 4 lb. 7.6 oz. throughout the summer season, whereas adult hens averaged 2 lb. 9.3 oz. for the same period. By September 1, the weights of juvenile males were comparable with the weights of adult females. By the end of October in Eden Valley, immature birds of both sexes had reached the approximate weight levels of adult birds and could not be distinguished from adults except by experienced field men. As a rule, hunters are highly selective when hunting sage grouse in midsummer and concentrate their gunning efforts on young birds and adult hens. Fall hunting seasons would produce a better proportioned harvest of the sexes and age groups at a time when the flocks are more widely distributed over their ranges.

There are other important considerations favoring autumn seasons. In August adult birds are at the height of their moult and in their poorest condition of the year. The preservation of birds killed during high temperatures of August is a problem. During the past forty-four years the average maximum daily temperature for August in Eden Valley has been 80.6° F. High summer temperatures prevent efficient work by bird dogs, serving to increase crippling losses. Summer seasons encourage hunting from automobiles due to the discomforts of walking in hot, dry weather. During the summer months sage grouse are concentrated in areas adjacent to open water. Once established, fall seasons would prove as effective as midsummer seasons in reducing damage pressure from sage grouse in ranching areas. Many western game and fish departments rely heavily upon the sale of nonresident licenses for major sources of their operating revenue. Fall seasons on sage grouse would enhance this policy by permitting nonresident hunters to hunt sage chickens as well as big game.

In view of the foregoing factors, it is recommended that sage grouse hunting seasons in the mountainous western states be held not earlier than September 20.

SAGE GROUSE POPULATIONS
AND LAND-UTILIZATION PATTERNS

Fluctuations in Population Levels

Invasion and Occupation of Sage Grouse Range by Man

One hundred and fifty years ago aboriginal Indian tribe roamed over the western prairies, hunted and fished these primi tive lands, and performed the "dance of the prairie chicken.' Since the intrepid John Colter made his explorations over th western plains and mountains, the fur trapper, settler, railroad er, stockman, reclamationist, and industrialist have in successio been lured onto western range lands, each attempting to improv on the system of land use and the pattern of life, long practice by the Indian, wildlife, and nature.

The history of our early land-use programs made the declin of native wildlife populations almost inevitable. Use of th range by certain wild animals, such as the American bison, wa incompatible with the rapidly developing system of higher lan uses; hence they were exterminated from the western range Game was held to be inexhaustible and along with other natura resources was exploited to the utmost. The species were lavishl utilized as a source of food, sport, amusement, and wealth, and in many areas they were completely displaced from the scene. In other districts their numbers were so reduced that for years their status was precariously uncertain. Such was the trend in game numbers and the pattern of game policy in the West well into the twentieth century.

With the creation of the federal land-management agencies, state conservation departments, and national, state, and private conservation societies, the period of exploitation of natural resources came slowly to an end. Wise use under proper management became the policy as most of the depleted forest, range, and wildlife resources were gradually restored or improved under a condition of permanency and sustained yield.

The pattern of decline in sage grouse numbers has been little different from that exhibited by numerous other game animals in the West. Destruction of habitat and inadequate protection, whatever their nature, have been basic causes of sage grouse decrease throughout the West as a whole. These decimating factors usually have operated through the media of early and extended shooting seasons, high bag limits, and overgrazing of public and private ranges, as well as elimination of sagebrush areas through agricultural practices and urban development. The oft mentioned decimating factors of unfavorable weather, increased predation, and disease may have been of significance in localized areas, but were relatively unimportant in the overall decline in sage grouse numbers.

Grouse Numbers and Human Economies

At the present time there is a striking correlation between sage grouse abundance in the various western states and corresponding human populations. It is more than mere coincidence that the states supporting the lowest human population densities contain the greatest acreages of sagebrush habitat and the largest sage grouse populations (Table 41).

States with large sage grouse populations have tremendous acreages in public ownership and have human population densities of substantially less than ten persons per square mile. Wyoming, for example, has 52 per cent of its land area in the public

TABLE 41

Relationship Between Human Population Densities and
Sage Grouse Abundance in the Western States

State	Human Population 1950 Census	*Relative Rank in Sage Grouse Production	Sq. Miles Habitat
Nevada	160,083	1	60,000
Wyoming	288,800	2	45,000
Idaho	588,637	4	30,000
Montana	591,024	3	40,000
Utah	688,862	5	20,000
Colorado	1,325,089	7	10,000
Oregon	1,521,341	6	15,000
Washington	2,378,963	8	8,000
California	10,586,223	9	2,500

*Based on square miles of habitat available to sage grouse

domain and a population density of about three persons per square mile. Almost 17 million acres in Wyoming are included in Taylor Grazing districts administered by the Bureau of Land Management. The bulk of this acreage is classified as sage grouse habitat.

The assignment of rank to the states in sage grouse production was based on a classification of sage grouse habitat and relative abundance of the birds as shown on preliminary sage grouse distribution maps furnished by all the western states, and not on actual grouse population figures. The latter method would have been more desirable, but such statistics were not available.

The growth and development of an agrarian mountain state like Wyoming is reflected in Table 42. A grazing and livestock industry has evolved as the primary economy of the state. In a land-use planning program the livestock industry can be considered the foremost activity in Wyoming. Public and private land-management agencies now generally recognize the production of wildlife to be one of the most important secondary uses of wild lands in Wyoming. Land management agencies have entered into co-operative agreements with the Game and Fish Commission and the sportsmen regarding the management of the wildlife resources in Wyoming. In as much as the continued prosperity of the livestock industry depends upon the preservation of tremendously large acreages of native forage types as well as upon the improvement of depleted ranges, it follows that sage grouse, antelope, deer, and other game animals which inhabit livestock ranges will increase simultaneously with livestock and prosper in response to good land management. Under a co-operative management plan, there is less likelihood of conflict between wildlife interests and human economies.

In areas featured by expanding human populations and agricultural development, it can be expected that sage grouse numbers will decrease mainly from elimination of their habitat. Sage grouse populations in the upper Green River Basin during the past thirty years have neither been jeopardized by an expanding human population nor by an alteration of basic land-use policies. The population census for the four counties of the Basin has indicated that, in relation to the present land-use policies, a condition of stability in population growth has been attained. On the basis of total land area, there are approximately two persons per square mile in this area which is considered to be the finest sage grouse range on the continent (Table 43).

TABLE 42

*Growth of Wyoming from 1870 to 1950**

Year	Population
1870	9,118
1880	20,789
1890	62,555
1900	92,531
1910	145,965
1920	194,402
1930	225,565
1940	250,745
1950	288,800

*Data from U. S. Bureau of Census

Effects of Land-Use Activities on Sage Grouse

Industrial Development

In spite of abundant supplies of natural resources, there has been relatively little industrial development in the western mountain states. Lack of development has been attributed to a shortage of electrical power, a critical shortage of water for industrial use, differential freight rates in marketing western products, severe winter climates, small numbers of people, and general dominance in the industrial field by the politically powerful eastern states.

The struggle for industrial supremacy has long featured the political and economic scenes of the normally agrarian western states. Political parties measure progress in terms of new industrial plants and defense projects acquired within their respective domains. Seldom is any consideration given to the detrimental effects of such development upon the wildlife, recreational, and esthetic values, which in most western states outrank all other industries in the state in terms of gross income, capital value, and participation by the general public. In view of a rapidly expanding population in industrial states such as California, Texas, Michigan, and Ohio, it would seem that states with low populations and superb recreational facilities could enhance local as well as national welfare by devoting a larger share of their resources and efforts to recreational development, rather

TABLE 43

Human Populations and Land Areas by Counties in the Upper Green River Basin from 1920 to 1950

County	Population				*Total Land Area
	1920	1930	1940	1950	
Lincoln	12,487	10,894	10,286	8,973	2,625,000 acres
Sublette	1,700	1,944	2,778	2,450	3,121,000 acres
Sweetwater	13,640	18,165	19,407	21,938	6,715,000 acres
Uinta	6,611	6,572	7,223	7,295	1,325,000 acres
	34,438	37,575	39,694	40,656	13,786,000 acres

*Includes 2,066,000 acres in national forests

than to an expanding industrial program and ultimate depletion of their expendable natural resources.

With the exception of its oil and coal industries, the economy of Wyoming is essentially that of an agrarian state. The raising of livestock and the related agricultural pursuits constitute the most important industry of the state.

Wyoming is currently experiencing considerable activity in its oil fields. Many of the existing fields as well as the sites of some new wells are located in sagebrush areas and conflict with sage grouse populations, mainly through the elimination of habitat and invasion of native ranges by industrial activity. Regulations governing the drilling and pumping of oil wells have partially eliminated stream pollution and unnecessary destruction of native vegetation. Abandoned wells must be sealed to prevent leakage into stream systems. All pools constructed for collection of waste material have to be filled in prior to abandoning the well site. Leveling operations generally restore the original contour of the land in the vicinity of the well site. The bulk of Wyoming's crude oil supply is transported into nearby states for processing, thereby eliminating the industrial activity associated with the refining and distillation processes.

Up to the present time oil activity has been of limited nature on the better sage grouse ranges in Wyoming. Numerous wildcat wells have been drilled in the Green River Basin, but practically all of these have been abandoned, or are not producing. One such well was drilled in 1948-49 by the Superior Oil Company on the Dry Sandy—Pacific Creek sage grouse study area. This well was drilled to a record depth of 22,000 feet with one of the largest rigs then in operation. The well was sealed and abandoned in 1949 with only slight alteration in the original aspect of the area. Studies conducted in the close vicinity of this well revealed that the drilling operations had not interfered seriously with sage grouse nesting and strutting activities. The birds continued to occupy areas adjacent to the well site throughout the drilling period, even to the extent of nesting within a few hundred yards of the rig, strutting within 500 yards of the operations, and watering daily at sediment pools located at the base of the rig.

The largest natural gas and coal deposits west of the Mississippi River are found in Wyoming. The present development of these industries has not materially altered the surface aspect of the land, although in the vicinity of mine workings some pollution of streams and the landscape has occurred from waste material.

The increased costs of mining deep deposits of coal, combined with a high competitive market from other fuels, has contributed to the development of the strip-mining industry in certain sections of Wyoming. Although former mining methods had relatively little effect upon sage grouse, strip-mining operations impose serious threats to wildlife populations, since vast areas of native vegetation and surface soils are destroyed in order to expose coal veins lying from 50 to 100 feet underground.

Many states have long been plagued with the destruction of thousands of acres of valuable farm and timber lands as an aftermath of coal stripping operations. The late Harold L. Ickes, former Secretary of the Interior, had much to do with coal mining during the war, and commented at length on coal stripping in a syndicated column appearing in July, 1948:

> In 22 states, including Ohio, Illinois, Pennsylvania and Kansas, there are great and growing areas of land, the surface of which, in effect, has been destroyed. While these sections are not yet as extensive or impressive as the famed Bad Lands of the Dakotas, their terrain is just as formidable and even more worthless. They do not even have the saving grace of natural interest. They are man-made, some of them from the finest farm lands.
>
> These truly bad lands are the result of the strip mining operations of men digging coal. A stripping operation in the coal fields gets its name from the fact that the coal lies close enough to the surface so that it does not have to be mined out through pits or tunnels. It can be scooped up by steam (or electric) shovels, some of which are said to have a capacity of 35 cubic yards at a single bite. The last cuts these monsters make are often 60 to 70 feet in depth and 100 feet or more in width, creating a huge chasm. These huge piles of uptorn earth are called spoil banks.
>
> There are many thousands of Americans who love nature and who sometimes even cry aloud over flagrant instances of

its devastation. We have active and devoted conservationists who have fought to establish and maintain our great national parks and monuments and who watch with a jealous eye the cutting down of any of our last remaining stands of magnificent virgin timber. But the type of destruction and defacement in the extensive strip mining areas of the country has attracted little attention and less protest. The least that the operating coal companies could do would be to take off the top soil and replace it as top soil after leveling the spoil banks at the end of the mining operation. This would add somewhat to the cost of the coal, but it would not cost anywhere nearly so much as the lost productivity of the despoiled land.

Under continued public pressure several eastern states, notably Ohio, Indiana, West Virginia, Pennsylvania, and Illinois have recently enacted strip-mining laws requiring coal operators to reclaim and improve stripped lands by leveling, preserving top soil, and planting to pasture-grasses and trees.

Coal stripping operations in the Green River Basin of Wyoming are being conducted at separate locations near Kemmerer and Rock Springs. It would be well for the citizenry of the state to evaluate possible deleterious effects of strip mining upon livestock and game ranges, and give worthy consideration to the enactment of strip mining laws before thousands of acres of valuable range lands have been destroyed.

A prospective shale oil industry may also present a threat to the survival of sage grouse through the elimination of sagebrush and pollution of streams. This will depend largely upon the nature of the mining and processing methods which are adopted. The Bureau of Mines has estimated that there are 300 billion barrels of oil in shale deposits in the United States. The majority of these shale reserves are located in the Rocky Mountain area, with large deposits in the Green River and Red Desert sections of Wyoming. The maximum yield in the Wyoming shale fields has been estimated at 35 gallons per ton of shale and 100 thousand gallons per square mile. There is no organized shale oil industry yet, but since the cost of obtaining oil from shale is only slightly more than the cost of refining petroleum, it has been predicted that the industry would begin to develop by 1960.

Agricultural Development

The rush of prospectors westward to gold fields during the mid-1850's launched the invasion of western lands by pioneers and early settlers. From the very outset, the favored homestead sites were located at the outlets of the mountain valleys and along the broad, alluvial, outwash fans and flood plains of major river systems. Here were to be found the most fertile soils and the most reliable supplies of water for establishing ranches and setting up livestock operations. Livestock roamed over adjacent semiarid sagebrush and short grass ranges.

In primitive times sage grouse populations attained their greatest abundance in the mountain foothills and along the fertile sagebrush-covered flats bordering mountain streams. It has been estimated that more than 50 per cent of the original sage grouse range has been destroyed by settlement and cultivation. Accompanying the settlement and cultivation of sage grouse ranges has been a maximum utilization of surface water supplies for irrigation and other domestic purposes. Entire streams frequently have been diverted from their channels with the resultant drying of stream beds as they moved onto the semiarid deserts. In many other instances the application of water to lands unsuitable for cultivation has resulted in severe leaching of alkali salts with resultant pollution of stream courses and conversion of fertile sagebrush lands into alkali-impregnated tracts featuring the salt shrubs.

As late as 1900, Big Sandy and Little Sandy creeks reportedly furnished excellent trout fishing in Eden Valley. However, with the creation of the Eden Valley irrigation district in 1908, and diversion of the runoff from these two streams for irrigation purposes, the trout rapidly disappeared, forcing local fishermen to retire to mountainous areas for their sport. Sage grouse were likewise affected by these land-use practices, and responded either by withdrawing to undisturbed areas or by decreasing to lower levels of abundance in the vicinity of the reduced and polluted water supply.

Stream diversion has continued to be one of the most important factors affecting sage grouse abundance throughout semi-arid regions of the West. Other game species have also suffered drastic reductions in numbers as a result of stream diversion. Waterfowl losses from botulism increase sharply in certain areas where large amounts of water are diverted for domestic purposes. This has been particularly true in the marshes bordering the Great Salt Lake.

In a previous chapter, instances were noted where sage grouse populations have been heavily depleted in Wyoming as a result of intensive agricultural development. There are many such areas in other western states. An excellent example of sage grouse displacement by agricultural activities is afforded along the base of the Wasatch Front in Utah. There are few tracts of sagebrush remaining on the foothill slopes draining into the Great Salt Lake Basin. Ranches, cities, and industrial locations dominate the entire landscape for a distance of perhaps 200 miles along the eastern borders of the Great Basin.

Similarly, agricultural development has eliminated much of the sage grouse habitat along the lower Snake River Valley in Idaho. On land areas in the Yampa River drainage in northwestern Colorado, heavy overgrazing and wheat-farming activities have combined to seriously deplete the sagebrush-grass ranges and reduce sage grouse populations. Considerable effort is now being expended by land-management agencies to restore this land to a condition favoring increased productivity for both livestock and wildlife. On depleted western ranges maximum utilization by sage grouse will be dependent upon not only a restoration of grass resources but also upon restoration and preservation of sagebrush—the most important component of sage grouse habitat.

The elimination of sagebrush ranges along lower river drainages frequently has little effect upon the spring and summer activities of sage grouse; yet quite often it completely destroys vital winter ranges. The species is extremely mobile, migrating from the higher summer ranges to ranges several thousand feet lower during winters with a heavy snowfall. The reduction in sage grouse numbers in many of these intensively farmed areas has been the direct result of elimination of the birds' winter range. This condition is becoming more widespread as existing agricultural areas are enlarged and as new areas are developed for homesteading.

Reclamation Programs

Western reclamation is now in "the high tide of its greatest program." According to reclamationists, the year 1950 was one of the greatest for western reclamation. Ten giant new dams started rising in six states of the Rocky Mountain Empire. The new dams will store irrigation water and furnish electric power for several hundred thousand acres of ranch and farm lands.

At the present time in the state of Wyoming there are four major reclamation projects and several smaller projects in operation. These projects will ultimately irrigate more than 500,000 acres of semiarid lands. Many additional projects are now in the survey, planning, and construction stages. In the upper Green River Basin alone, nine reclamation projects have been proposed along the tributaries and the main channel of the Green River. To allow for silting, the Bureau of Reclamation is increasing the necessary water storage capacity of many reservoirs by 100 per cent, thereby doubling the effective life of the storage system, but at the same time decreasing still further the amount of habitat available for terrestrial wildlife.

The welfare of wildlife will be affected by virtually all of the western reclamation projects. On the Riverton project in Wyoming, which embraces almost 100,000 acres of land originally covered with sagebrush, the elimination of habitat has reduced the original populations of sage grouse to the point of extirpation in a period of twenty years. In many cases native wildlife species have been displaced by exotic forms. High pheasant populations and excellent pheasant hunting have been characteristic of reclamation projects lying at elevations below 5,000 feet. Waterfowl have increased noticeably in the vicinity of most irrigation projects. The presence of small and isolated projects of from 10 to 20 thousand acres in semiarid sagebrush regions has created conditions favorable to the concentrating of sage grouse populations as is evidenced in the Eden Valley irrigation district.

Many of the proposed sites for reclamation projects are valuable breeding and winter ranges for big game as well as sage grouse. Still others will act as barriers to the seasonal movements of both game and fish, either by the location of ranches and fences, dam construction, or the drying of stream beds below water impoundments.

Considerable investigation is being conducted by the Office of River Basin Studies, U. S. Fish and Wildlife Service, on the different reclamation areas proposed for development in the West, but for various reasons the interests of sportsmen and the welfare of wildlife do not always receive proper attention in the plans for development of many western reclamation projects.

Wildlife values are difficult to express in monetary terms, and when analyzed solely on that basis, can hardly compare with the high values placed upon the accrued benefits from electric power and irrigation water. Under our present national and state land-planning programs, the interests and policies of the various agencies managing wild lands are usually subservient to those of powerful reclamation agencies.

The Office of River Basin Studies is charged with the responsibility of evaluating the fish and wildlife resources in relation to plans for river basin development. The worth of wildlife is determined on the basis of an arbitrarily established monetary value assigned only to economically important animals. The total net gain or loss is fixed in relation to an animal's abundance before project development and after project development. Fish and Wildlife Service biologists conduct brief surveys of the project area and its fish and wildlife populations prior to final approval of the project. Quite often little study is given to seasonal movements of fish and game as they relate to utilization of the project area throughout the year. Conflicts between wildlife and agricultural interests, inevitably created when irrigation projects are thrust into extensive game range, are seldom resolved in these evaluation reports. As far as the states are concerned, the Fish and Wildlife Service evaluation reports too often are submitted to the Secretary of the Interior without adequate consideration being given to viewpoints, policies, and recommendations of the state game and fish departments.

The sage grouse is one of the most abundant game species in the upper Green River Basin, and will be vitally affected by developments currently being proposed by the Bureau of Reclamation in that area. The proposed Seedskadee project is designed to provide irrigation water for 60,720 acres of sagebrush land located along the Green River in Sweetwater and Sublette counties. In a preliminary evaluation report on the relation of fish and wildlife resources to this project, the Fish and Wildlife Service has estimated that there is a sage hen population of one bird to 15 acres in the project area with an annual value of $4,050. They consider that there would be a net reduction in upland game values as a result of the project. They state that there, no doubt, would be a decrease in the sage hen population, but this is partially justified by the Service since the cultivated areas *should provide* some habitat for pheasants and Hungarian partridge (at elevations approaching 7,000 feet). Under project conditions they predict that upland game would have an estimated annual value of $2,530, to give a total annual evaluation loss of $1,520 (Table 44). Their bioloigsts believe that it might be possible to help the wildlife populations remaining under project conditions by exercising control of grazing, protection of watersheds, control of predators, and improvement and more careful management of antelope and sage hen habitats near the project area.

In the preliminary evaluation report a value of $1.00 per bird has been assigned to the resident sage grouse population on the Seedskadee project. No allowance has been made for the thousands of sage grouse which annually winter on lands included within the proposed development area. Neither has there been a provision for damage control payments on sage grouse populations likely to inflict damage to domestic hay crops on the new project. Under actual conditions it is more likely that sage grouse numbers in the vicinity of the new project will have to be reduced to lower levels of abundance in order to suppress the damage potential.

TABLE 44

*A Summary of Wildlife Values on the Seedskadee Reclamation Project**

Wildlife	Without-the-Project Values	With-the-Project Values	Net Gain or Loss
Big Game	$ 4,700	$————	—$ 4,700
Upland Game	4,050	2,530	— 1,520
Fur Animals	1,690	1,560	— 130
Waterfowl	200	230	+ 30
	$10,640	$ 4,320	—$ 6,320

*Taken from a preliminary evaluation report on fish and wildlife resources in relation to the Seedskadee project—Office of River Basin Studies, U. S. Fish and Wildlife Service.

The nation has embarked upon a gigantic and costly program of dam construction on practically every major and minor watershed in the country. Projects are frequently approved and completed with little justification from the standpoint of local and national economy, the reports of the various governmental bureaus notwithstanding.

The Eden Valley Reclamation project in southwestern Wyoming has been in existence for more than forty years. The present development was initiated under terms of the Carey Act by the Eden Irrigation and Land Company in 1907. The project apparently represented an attempt on the part of the government to provide homesteads for families which had been displaced from submarginal agricultural areas of the Middle West. The project was located in a semiarid region at an elevation higher than 6,500 feet. The winters are long and severe, the growing seasons unpredictable and short at the best, the soils of questionable quality, and the water supplies for potable purposes extremely unsatisfactory. At the period of maximum development a total of some 131 farm units of approximately one-quarter section size (160 acres) were taken up by homesteaders. At the present time the number of landowners has dropped to 84, representing a 36 per cent decrease in landownership. Twenty of the remaining landownership units have either reverted to the county in lieu of taxes, or else are being operated unprofitably. On the basis of economical operation approximately 50 per cent of the original homestead units have proved either incapable of supporting their occupants or the occupants have been incapable of managing their homesteads. Further study of the landownership pattern reveals that a few families now control the majority of Eden Valley lands. This is not surprising since from 1935 to 1945 the consolidation of farms and ranches in Wyoming occurred at a rapid rate; many of them proved to be too small for economical operation. From 1920 to 1945, the average size of a Wyoming ranch increased from 750 acres to 2,532 acres.

As a result of poor design, faulty construction, and inadequate maintenance, the irrigation structures had deteriorated to such an extent by 1950 that continuous delivery of irrigation water was jeopardized. Future operation of the Eden Valley project was dependent upon extensive repair and rehabilitation work beyond the financial capabilities of local water users. Congress, thereupon, authorized this repair work in 1950 and *further provided* for the construction of a new dam on the Big Sandy Creek at a cost of $6,000,000; included was a provision to enlarge and extend the present irrigation system to serve 20,000 acres of land instead of 9,000 acres. Local water users were obligated to repay approximately 25 per cent of the construction costs; the balance was to be paid by prospective industrial plants in the upper Colorado River Basin, or be considered nonreimbursable.

This new development on the Eden Valley project has been undertaken with a full knowledge of the many hazards to a sound agricultural economy. Although the project has been blessed with abundant aid from the federal treasury and a most competant staff of engineers, soil scientists, and construction personnel, the fact remains that the suitability of the area for agricultural development was no better in 1950 than it was in 1907.

The climate has not changed; the soil and drainage conditions have not improved; the supply of potable water remains critical; and increasingly larger units of land are required to maintain an acceptable standard of living. The area is so isolated that rural electrification has not been available, although fully 94 per cent of the rural areas of the nation are now served by an electrification program. The agricultural opportunities on the project are such that a large percentage of the landowners and tenants have accepted full time local employment with the Bureau of Reclamation and the Soil Conservation Service. Others work from 40 to 60 miles away in the industrial towns of Rock Springs and Green River during the winter months. The system of granting submarginal homestead sites to veterans of low financial means practically assures a rapid turnover in land-ownership and eventual dominance in ownership by a few large operators.

On thousands of semiarid acres in the West, the application of irrigation water has created conditions favorable to the development of alkaline soils and vegetation. This has been particularly evident in many of the areas reclaimed by reclamation agencies. Large tracts of land supporting rich surface-soils and sagebrush-grass vegetative types are underlain with impervious rock formations of marine origin. Many areas cannot be irrigated without the risk of abandonment due to alkali accumulations on the surface. Other lands require elaborate and expensive drainage systems to prevent undesirable salts from rising to the surface layers, or else require a most careful application of water by irrigators. Seepage emanating from irrigation structures has often created alkaline conditions on lands adjacent to areas of cultivation. In addition to the abandonment of farm lands, the creation of alkaline soils and the accompanying salt shrub vegetation has lowered grazing values, and reduced livestock and game carrying capacities.

Sizeable tracts of land in the Farson segment of the Eden Valley irrigation district have reverted to alkali as a result of conditions previously described in the soils section. The Eden segment of the district is characterized by relatively minor tracts of alkali land. The high incidence of alkaline soils and salt shrub vegetation in the Farson segment has profoundly affected the degree of use and occupancy of that region by sage grouse.

As has been noted elsewhere in this work, average sage grouse nesting densities on the Farson segment during 1949 and 1950 were thirty-four nests per square mile as compared with forty-eight nests per square mile in the Eden segment for the same period. Censuses of males on strutting grounds revealed an average of six cocks per square mile on the Farson segment as compared with eighteen cocks per square mile on the Eden segment. Similarly, alfalfa fields in the Eden segment sustained far more damage from sage grouse than did alfalfa fields in the Farson segment. Ninety-five per cent of some three thousand birds trapped during 1949 and 1950 were removed from ranches in the Eden or southern segment of the district.

It is no coincidence that the bulk of the land selected for homesteading during the current expansion of the Eden Valley project will be located on the Eden segment. Those lands supporting the best soils as well as the highest numbers of sage grouse have been selected for agricultural development, according to preliminary planning reports received from the Bureau of Reclamation and the Soil Conservation Service.

The debates in Congress and in the nation's high tribunals attest to the constant struggle for water rights in the West. Interstate compacts allocate the total water flow from the river systems of the western states. There is a growing feeling among reclamation agencies that every spare drop of water on western watersheds must be diverted for use in industrial, agricultural, or electrical power projects. In a great organized effort to accomplish these objectives, major river systems are being diverted from one watershed to another, high mountain lakes are being tapped, and plans are being made to freshen sea water; even the clouds are being exploited by rain makers! None of these vast programs have given much consideration to the use of water by the recreation industry or wildlife, except as it may be made available to them within the pattern of the broad objectives.

In their search for water to serve industry, agriculture, and urban areas, the Bureau of Reclamation is sending their engineers higher and higher on the western watersheds. In the upper Green River Basin, existing and proposed projects already extend into the mountain foothills of the Wind River, Wyoming, and Uinta ranges. Numerous high mountain lakes have already been tapped in these ranges to meet the water demands of reclamation projects. Others are included in schemes to increase water storage capacities by means of engineering works at their outlets. Encouraged by some of their earlier successes, reclamation officials have predicted that 50 million acres, west of the Rockies alone, can be watered into agricultural production through irrigation.

One need only study the history of agricultural development on the treeless Great Plains to realize the great risks involved in attempting to completely develop the semiarid regions of the West for a prosperous agricultural economy, and at the same time sacrificing an old and soundly established grazing economy —a policy which would ultimately eliminate those domestic and game animals dependent upon undisturbed sagebrush lands for their survival.

So vast and complicated have become the nation's water problems that the President in January 1950 appointed a Water Resources Policy Commission to make an intensive investigation. The Commission found that there are staggering problems in the West where the great bulk of the billions of dollars appropriated to conserve soil and water has been spent. The program is out of balance. Too much money has been expended on big river improvements and multiple-purpose dams, and too little allocated to restoring the abused watersheds, the depleted forests, and the overgrazed ranges. The Commission emphasized that the wildlife resources, especially fisheries, have largely been neglected in the river basin programs. Scientific studies have indicated that water supplies on many watersheds are insufficient to meet all the demands proposed in reclamation projects. There is great danger in placing people on lands which may not sustain them indefinitely. The Commission summarized its study with the conclusion that the United States has been dealing with its water problems in a piecemeal, irrational, and wasteful manner. Instead of a co-ordinated approach that would yield the best results, the nation has planned its public works too often on the basis of short-term economic expediency and sectional pressures. Moreover, the country has paid excessive attention to engineering works on the river systems, neglecting the deteriorated lands which drain into the rivers and the wild and domestic animals which live on the lands. According to the President's Water Resources Policy Commission, continuation of present policies, or lack of policies, will mean a continuing waste of money and effort in the pursuit of conflicting goals.

Livestock Industry

Overgrazing and drought have created range conditions adversely affecting the relative abundance and distribution of sage grouse populations. Within recent years these influences have exerted less effect upon sage grouse as a result of range management practices of the livestock industry and other regulatory agencies. However, the effects of intensive range use by domestic livestock during the era of complete exploitation should not be underestimated.

As early as 1906, sportsmen and conservationists expressed concern regarding the downward trend in sage grouse populations in Wyoming and other western states. In a few areas adjacent to the mountains in Wyoming it was reported that sage grouse were maintaining their numbers, but over the state as a whole the trend was definitely downward, as is attested by the closing of the hunting season in Natrona and Sheridan counties in 1909. At this period in the development of the West, neither illegal hunting nor the automobile could be considered directly responsible for initiating the destruction of sage grouse populations. It is a principle of wildlife management that major trends in game abundance over extensive range can be traced to some significant alteration in the pattern of land use. As pointed out, the decline of sage grouse populations has not been of recent occurrence, but, on the contrary, it coincided rather closely with the period of maximum utilization of range resources by livestock from 1900 to 1915. It was during this period (1909) that the number of stock sheep on Wyoming range lands topped the six million mark. There apparently are no accurate records of the additional thousands of sheep which were trailed across Wyoming from range lands in states to the westward en route to markets in Kansas and Nebraska. Since that period sheep populations in Wyoming and other western states have declined steadily. Sheep numbers in Wyoming reached a low of 1,863,000 head in January, 1951. From a high of 220,000 head in

1919, horse numbers (exclusive of unbranded stock) have declined to a record low of 76,000 in 1951. Annual roundups of unbranded stock have all but eliminated the wild horse from the Wyoming range. In 1919 cattle numbers in Wyoming reached a maximum of 1,301,000 head. Since that time they have declined appreciably, although within the past decade their numbers in Wyoming have been fluctuating at the one million mark. Increasing costs of production, a scarcity of dependable labor, and other economic factors have caused many sheep producers to reduce the size of their operations in recent years. Many woolgrowers have switched from sheep to cattle operations. The trend in numbers of livestock on Wyoming ranches and farms from 1870 to 1950 is shown at five-year intervals in Table 45.

TABLE 45

*Number of Livestock on Wyoming Range Lands from 1870 to 1951**

Year	Class of Livestock (Unit=1,000 Head)		
	Sheep	Cattle	Horses
1870	36	71	1
1875	256	218	5
1880	517	523	12
1885	793	779	30
1890	1,431	700	95
1895	1,718	625	120
1900	3,675	611	128
1905	3,547	750	133
1909	6,023	765	153
1910	5,480	746	160
1915	3,633	887	190
1919	3,602	1,301	220
1920	2,960	950	220
1925	2,613	795	200
1930	3,420	790	176
1935	3,444	858	146
1940	3,778	811	128
1945	3,040	1,043	118
1950	1,901	1,001	80
1951	1,863	1,141	76

*Data furnished by the Bureau of Agricultural Economics, U. S. Dept. of Agriculture.

The Mountain West is the center of the western range livestock and grazing industry. The sagebrush-grass and salt-desert shrub range types are largely unappropriated public domain. These plant types provide the main habitat for sage grouse, antelope, deer, and a host of other wildlife species. Supplemented by the forest ranges, these types furnish the major range requirements for millions of sheep, cattle, horses, and goats. Throughout its northern extension, the sagebrush-grass type forms an indispensable link, both in spring and fall, between summer and winter ranges for cattle and sheep. The salt-desert shrub type is the winter home for several million sheep and several hundred thousand cattle.

The ownership and control of western range lands is complex. The grazing policies are established by federal, state, and private land-management agencies. Range operations in Wyoming are under the control of several federal agencies. Many range policies exert an adverse effect upon sage grouse and other wildlife species associated with the major range types. Other range practices are a distinct benefit to the welfare of these same game animals.

For many years a private sheep association has controlled the grazing rights on a 20-mile strip adjacent to each side of the Union Pacific railroad right of way in southwestern Wyoming. This area is primarily covered with sagebrush and the salt-desert shrubs and provides wintering range for sheep herds which belong to members of the association. Studies by Vass and Lang (op. cit.) revealed that the range lands immediately surrounding the controlled lease were subjected to more intensive grazing just prior to the opening of the winter lease and again following the closing date of the lease. This was caused by the sheep operators holding their bands on the border of the lease a few weeks prior to its opening date and again utilizing the same area between the time the lease had closed and the entry dates on the national forests. Plant-transect studies made on these areas in 1936 disclosed that the number of sagebrush plants which had died in one year on the land surrounding the winter lease was 30 per cent greater than the number of seedings. The number of sagebrush plants which had died in plots on sheep bedgrounds was

85 per cent greater than the number of seedlings. On plots within the winter lease and on spring and fall ranges near Eden Valley, the number of seedlings exceeded the number of sagebrush plants which had died.

Exclusive of livestock ranges controlled by the Forest Service, the U. S. Bureau of Land Management is the principal agency administering grazing use on public lands in the Green River Basin. For purposes of management the Basin lands have been divided into two grazing districts, namely: Wyoming grazing district No. 5, composed of 1,121,000 acres; and Wyoming grazing district No. 4, composed of 4,827,160 acres.

Grazing allotments for big game animals and livestock are established jointly by the Bureau of Land Management, the U. S. Forest Service, the Wyoming Game and Fish Commission, and the district advisory boards representing the livestock industry. These allotments are authorized and approved annually under the terms of the Taylor Grazing Act, legislation enacted in 1934 to rehabilitate western ranges and regulate grazing use for the sustained production of livestock and wildlife. Livestock and big game grazing allotments for the public grazing lands in the upper Green River Basin appear in Table 46.

TABLE 46

*Annual Range Allotments for Livestock and Big Game Animals in the Upper Green River Basin**

	Grazing Dist. No. 4	Grazing Dist. No. 5
Domestic Sheep	388,255	124,194
Cattle	33,433	53,749
Horses (licensed)	1,400	2,659
Horses (unlicensed)	1,000	-- ------
Mule Deer	6,500	3,610
Antelope	6,000	1,500
Elk	1,000	1,285
Moose	65	150

*Data from Bureau of Land Management, U. S. Dept. of Interior

Eden Valley homestead units have been collectively assigned livestock allotments for 1,575 sheep, 691 cattle, and 59 horses. This stock grazes range lands adjacent to the irrigation district. In the spring and again in the fall some 75,000 sheep were trailed across lands of the Eden Valley and Dry Sandy—Pacific Creek study areas en route to summer ranges in the mountains and winter ranges on the semiarid desert. The spring movement of these sheep herds across the study tracts commenced on or about April 15, and coincided very closely with the peak of sage grouse nesting activity. Some nest desertion was known to have occurred due to sheep movements across nesting areas, although under normal trailing operations of five miles per day, nesting mortality from this cause was not considered too serious. The effects of sheep-bedding activities and overgrazing on sage grouse nest mortality and on nest densities have previously been discussed.

John Hay, Jr., associated with the Blair-Hay Sheep Company, Rock Springs, Wyoming, has estimated that during the decade following the turn of the century approximately 800,000 sheep annually grazed range lands on what is now Grazing District No. 4. Just prior to the grazing reductions authorized by the Taylor Grazing Act, he has estimated that there were probably not less than 350,000 sheep on the Rock Springs Grazing Association lease. These numbers compare with slightly less than 400,000 sheep on grazing District No. 4 and about 190,000 sheep on the Rock Springs lease in 1951. Since the advent of the Taylor Grazing regulations there has been almost a 50 per cent reduction in sheep numbers on most grazing districts in Wyoming. It was during the period from 1935 to 1950 that the numbers of sage grouse, antelope, and mule deer began increasing to the phenomenally high levels of abundance which now exist in Wyoming and numerous other western states. This increase in game populations has been assisted by excellent protective measures and intensive predator control. However, the most fundamental factor causing this increase in game numbers has been an inspired and co-operative range restoration program, sanctioned and promoted by the livestock industry, by wildlife interests, and by the federal and state agencies engaged in the management of wild lands.

The majority of ranchers and stockmen have been extremely tolerant of the presence of maximum numbers of game animals on western ranges. This group has often acted as a desirable check upon sportsmen's organizations who have campaigned for game populations in excess of the allotted capacities of the range. Certain practices of the livestock industry have assisted in the management of wild game. Range improvements have served to effect a better distribution of livestock and game animals and to eliminate overuse of critical areas. A substantial portion of western range wildlife is produced, reared, and harvested on private lands of ranchers and stockmen. The restoration of depleted sagebrush range lands through controlled grazing and replanting of native grasses as *subdominant plants* in the vast sagebrush types has improved the habitat for sage chickens and increased grazing values for big game animals. Without the network of sheep wagon trails and graded roads constructed in the interests of the livestock industry, it would be difficult, indeed, to manage and annually harvest the increased numbers of sage grouse, antelope, mule deer, and elk on western range lands.

In the past there have been several attempts on the part of factions in the livestock industry to promulgate certain range policies which, according to the public viewpoint, were not always considered to be in the best national interests. The latest of these policies, promoted by the woolgrowers, is concerned with the erection of permanent type woven wire fences on Taylor grazing lands. In numerous cases the erection of these fences has been sanctioned by the Bureau of Land Management. The fences have been designed by the livestock industry to decrease the costs of sheep operations and to effect a better utilization of range forage. By the nature of their location and construction, they are destined to provide barriers impeding the normal seasonal movements of antelope herds. They will also impose hazards to sage grouse in the form of obstructions in their line of flight. Lastly, they will seriously interfere with the values of the hunt, either by preventing the freedom of movement of the hunters or of their quary, especially in the case of sage chickens and antelope. Substantial acreages of public range lands in Wyoming are now enclosed with woven wire fences which antelope are unable to go through or to jump under normal conditions. There is little merit in adopting a major range policy on public lands which is designed to jeopardize the welfare of either the game animals or the sporting public, which under the public land laws are guaranteed rights of access and use comparable to those exercised by the livestock industry.

Transportation Systems

The introduction of the automobile has been credited with causing serious inroads upon sage grouse populations. As early as 1916, game administrators in Wyoming were becoming alarmed over the opening of new districts to sage chicken hunting through the advent of the automobile. They found that cars were the source of much hunting in areas heretofore scarcely visited by hunters, and predicted disastrous results for sage grouse. When automobiles were not in general use, it required a long trip with a team in order to reach the districts which supported the largest numbers of birds. With the automobile as much as 100 to 200 miles could easily be traveled in a day in 1917, with the result that the former supply of birds had been greatly reduced. In those days sage grouse were legal game throughout most of the summer and fall periods. The advent of automobiles and road systems undoubtedly exposed sage grouse populations to undue hunting pressures, both legal and otherwise. This source of mortality has largely been eliminated as legal hunting seasons were shortened and as illegal hunting became difficult.

Large numbers of sage grouse are killed annually by vehicular traffic on highways which traverse their breeding areas. The construction of high speed highways across western prairie regions imposes additional hazards to wildlife populations.

Airports constructed in sagebrush areas often have been utilized by sage grouse as strutting areas. The local citizenry treasures most of these sites for the opportunity they afford of observing the birds' strutting displays. As has been stated, strutting sage grouse sometimes interfere with scheduled air operations on the airport at Riverton, Wyoming. Airline officials have periodically requested the game department to eliminate this group of birds, although a very slight effort on the part of airport attendants just prior to air operations would readily flush these birds from the runways. There has been a tendency for construction companies to wantonly clear sagebrush in order to build airstrips on lands adjacent to their projects. Airstrips constructed in this manner have been a feature of public lands near the Superior Oil Company's deep well on the Dry Sandy—Pacific Creek area and also adjacent to the Big Sandy Dam now under construction in Eden Valley.

Sagebrush Eradication Programs

The wide adaptability of sagebrush for survival under varying conditions of climate has given rise to a widespread belief that it is a noxious plant with low forage values, and should be eliminated throughout large areas of the West. According to range experts' comments in the 1948 Department of Agriculture Yearbook, sagebrush is considered to be of low forage value to livestock and is seldom browsed by sheep unless other forage is scarce or covered with snow. They considered the more palatable plants to be waning on the desert winter ranges and that sagebrush was invading the foothill ranges (Price and associates, 1948). Price (1948) has stated that on range areas now occupied by low value or noxious range plants such as mesquite and sagebrush, full returns from proper stocking, improved management and reseeding will not be possible until such soil-moisture "robbers" are eradicated, or at least controlled.

The prominent seed firm of Northrup, King and Company (1949) has publicized to stockmen that the sagebrush-grass ranges are badly in need of improvement, and that this can be accomplished either by giving the land a rest and allowing the native plants to regain their former prominence, or else by completely eliminating the sagebrush and reseeding to grasses.

The Bureau of Reclamation has gone to great length to put across the idea that western sagebrush lands are worthless unless they are reclaimed for agricultural purposes. A recent Bureau press release carried by the Associated Press stated that irrigation waters have been diverted onto hundreds of thousands of acres which once grew nothing more valuable than sagebrush. The edition of Time magazine for July 30, 1951, featured an extensive article on Bureau of Reclamation activities in the West.

Sagebrush land is described as mournful country, far more depressing than a self-respecting desert. A person lacking knowledge of America's western range lands might easily be impressed by the propaganda being put forth by the Bureau of Reclamation. Enthusiastic bureaumen have stated, "We and our contractors enjoy pushing rivers around." This declaration seems at times to aptly summarize their aspirations.

Federal agencies engaged in the management of wild lands recently have been conducting experiments on the various methods of eradicating sagebrush. These test-activities have included controlled burning, application of chemicals, and removal by means of mechanical implements. The sagebrush eradication program has gained prominence in the livestock economy as a result of several factors.

In recent years there has been a notable decrease in the numbers of sheep on western ranges, coupled with an increase of cattle numbers. This change in the ratio of sheep to cattle has complicated the forage relationships since the latter are predominantly grazing animals (grasseaters), whereas sheep utilize larger amounts of browse (woody plant parts) in their diets. Accelerated reclamation programs in the West have placed valuable range lands under cultivation and further reduced the amount of range available to livestock. Postwar economic recovery programs and the current national emergency have thrown increasingly heavy demands on the livestock industry. The industry in turn can be expected to demand increases in livestock allotments and an increased forage production. This situation has tended to upset the existing balance between livestock numbers and available range, or at least potentially so. The livestock carrying capacities of considerable areas of range are below the standards desired by many range managers and livestock operators.

The Bureau of Land Management has established a series of eradication and reseeding plots in the sagebrush-grass range type west of Eden Valley. Crested wheatgrass was planted on strips alternately cleared of sagebrush and left undisturbed. The survival of wheatgrass was negligible on the cleared strips. On the plots where the sagebrush was left undisturbed as a cover plant, the initial establishment of the grass was noticeably higher.

The reductions in livestock allotments currently being put into effect on some national forests have placed heavier demands upon the Bureau of Land Management to provide compensatory range on Taylor Grazing lands. The Bureau is engaged in experimental studies designed to eliminate sagebrush and establish grasslands on the foothill ranges adjacent to national forests. In this manner they hope to provide summer range for livestock being displaced from summer ranges on the forests. This particular program poses serious threats to the survival of sage grouse populations which habitually select these foothill ranges for their main nesting areas. The continued existence of big game herds is dependent upon the availability of woody plants on foothill ranges during the winter period. If grasses were to become established as the dominant vegetation on big game ranges, they would not be available in the winter due to snow conditions. Livestock are normally moved to wintering ranges at lower elevations and hence would not be directly affected by the elimination of browse plants in the foothills.

The agricultural conservation program of the Production and Marketing Administration, U. S. Department of Agriculture, contains provisions for making substantial payments to private landowners in return for clearing sagebrush and planting the land to pasture grasses or crops. This subsidy payment normally amounts to 50 per cent of the cost of clearing, not to exceed $5.00 per acre of land cleared. During the years from 1941 to 1949 approximately 24,000 acres of private land originally covered with sagebrush and associated browse plants have been converted into crop and pasture land in the four counties of the Green River Basin in accordance with the provisions of the agricultural conservation program. The total amount of land in private ownership in Sweetwater, Uinta, Sublette, and Lincoln counties is 3,978,000 acres.

In co-operation with the state agricultural extension services and the various implement companies, the U. S. Soil Conservation Service is actively supporting an extensive sagebrush removal program in the intermountain region. According to the Soil Conservation Service, millions of acres of range land in the intermountain area are today offering farmers and ranchers only 10 to 50 per cent of the range feed that they are capable of producing. Large areas formerly covered with palatable grasses and shrubs now support only sagebrush and cheatgrass. Ranchers are being advised by the Service to clear sagebrush lands and reseed to crested wheatgrass and other good grasses to improve their livestock yields and to gain a more effective control of erosion (Miles, 1946).

Nourished by liberal subsidies and support from the federal government and by influence from agricultural implement companies, the sagebrush removal and grass reseeding program is gaining favor among many ranchers and farmers as the cure-all for their range forage problems.

During a recent visit to the Great Divide Experimental Range in northwestern Colorado, I witnessed the demonstration of no less than ten implements for the control of sagebrush. Machinery designed for sagebrush removal included rotary cutters, brush mowers, root cutters, and disk plows. The demonstration was part of a range improvement field day sponsored by Colorado A & M College, the Bureau of Land Management, the Soil Conservation Service, and Moffat County ranchers and merchants.

As a result of overgrazing and wheat-farming activities, sage brush lands in western Colorado were badly deteriorated during the period following World War I. The improvement of depleted sagebrush lands in western Colorado has been a major activity of these agencies since 1941. Hull et al. (1950) have estimated that 4 million acres of land in western Colorado support sagebrush (*A. tridentata* and related species). According to these range men, much of this land can be improved by proper management and natural seeding, although on about 1 million acres the native forage plants are so sparse that artificial reseeding is needed to restore the plant cover within a reasonable time.

Most of the studies on forage yields on sagebrush ranges have been based on the amount of grass produced per acre, to the exclusion of browse plants. In many cases this method of evaluation has produced figures which inaccurately represent the total amount of consumable forage produced per acre. There is little question that the reseeding of the more productive sagebrush

Sage grouse breeding ranges in the mountain foothills are covered with deep snow in the winter

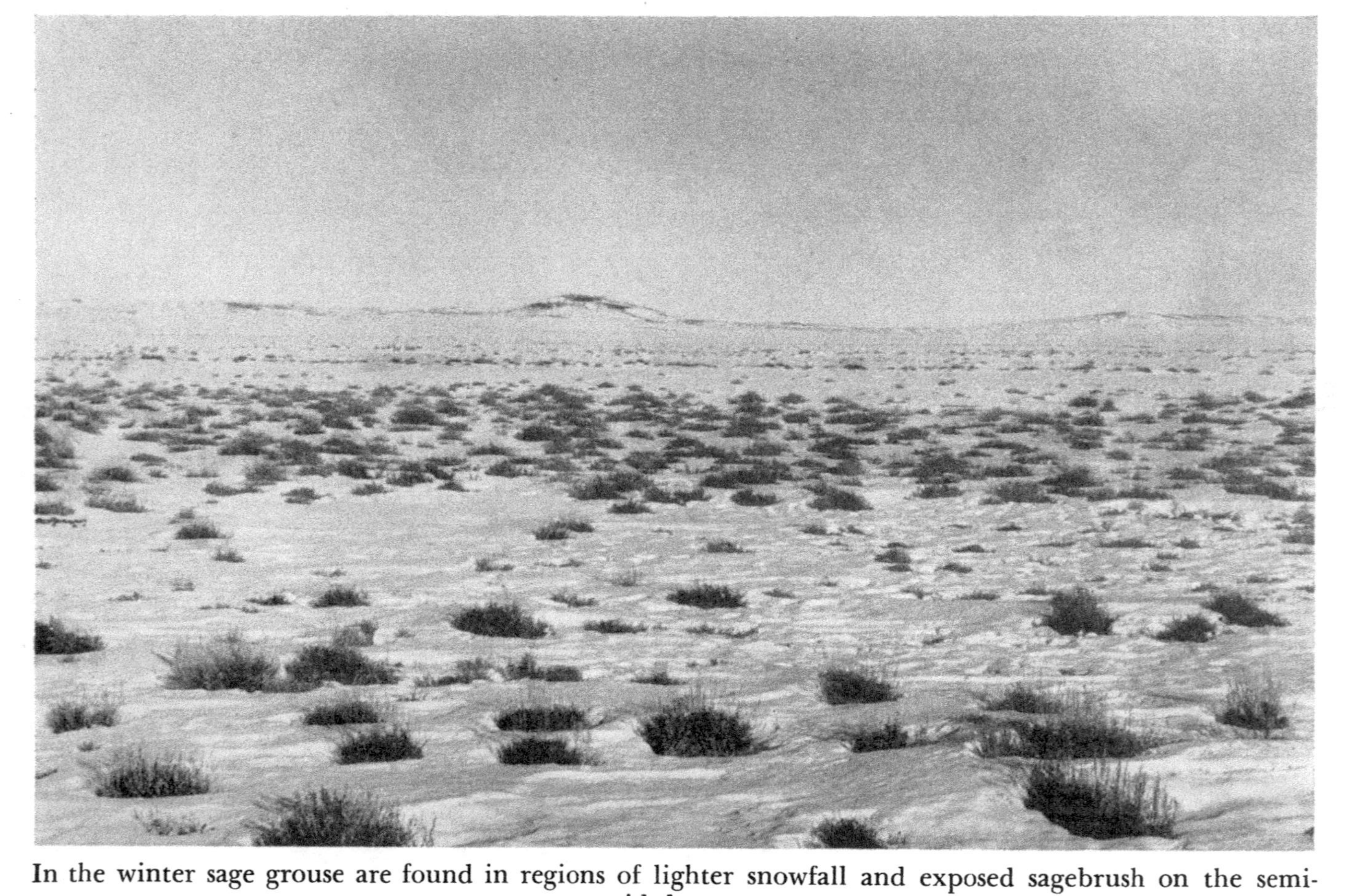

In the winter sage grouse are found in regions of lighter snowfall and exposed sagebrush on the semi-arid deserts

The vast majority of Wyoming's antelope herds are located on sagebrush-grass range types

Antelope share seasonal ranges with sage grouse in Wyoming

Sagebrush is probably the most important winter food of mule deer in the intermountain region

Cottontail rabbits rely heavily upon sagebrush types for their habitat requirements in many parts of the West

Several hundred thousand sheep annually trail across sage lands in the Green River Basin between summer ranges in the mountains and winter ranges on the deserts

Domestic sheep in the intermountain region depend upon sagebrush-grass types for their forage requirements from five to seven months out of the year

Sustained production of livestock as well as wildlife is vitally dependent upon preservation and careful management of sagebrush lands in the Mountain West

Grazing of livestock constitutes an important use of sagebrush lands in Wyoming

World's largest oil rig (228 feet high) and deepest well (22,000 feet) on Dry Sandy—Pacific Creek sage grouse study tract. *Photo by Standard Oil Co. (N. J.)*

Airstrip on **Dry Sandy—Pacific Creek** area. Airfield construction and associated operations are examples of land-use activities encroaching upon sage grouse range

Elimination of sagebrush habitat poses a constant threat to sage grouse survival in many parts of the birds' range. Sagebrush removal demonstration with Ferguson disc plow near Craig, Colorado

Demonstration area mowed in July, 1950, and drilled to crested wheatgrass in fall of 1950. Photo taken July 20, 1951. Range improvement practices which preserve sagebrush-grass types will aid sage grouse populations

Improvised mowing machine used in sagebrush eradication and control programs

The McCoy sweep designed to cut sagebrush roots

lands to pure grasses will initially increase the grass production —in many cases ten-fold. However, under the existing conditions of climate, soils, and elevation, there seems little likelihood of establishing and permanently maintaining a dominant grassland range type in the intermountain region. Without monetary assistance few ranchers, indeed, could afford the expense and great effort required to periodically clear sagebrush from their lands and reseed to grasses.

In a large sense, the sagebrush eradication program is being carried out in total ignorance of the tremendous value of browse plants to the welfare of livestock and wildlife. Any land-use practice which has as its objective the permanent elimination of sagebrush and establishment of grasses in the Mountain West will ultimately reduce the collective carrying capacity of that range for livestock (particularly sheep) , elk, mule deer, antelope, sage grouse, and many smaller game species.

It is noteworthy that the woolgrowers have not endorsed the program designed to eradicate sagebrush, the salt-desert shrubs, and other browse plants associated with these major range types. Innumerable sheepmen and sheepherders whom I have known in the upper Green River Basin have stated unequivocally that sagebrush and the salt-desert shrubs are utilized extensively by their sheep herds during all seasons of the year, irrespective of snow and forage conditions. The availability, palatability, and high nutritive content of sagebrush leaves combine to give the plant a high forage value.

Recent studies by Cook et al. (1951) on the consumption and digestibility of winter range plants by sheep have demonstrated the importance of browse plants in the sheep's diet. Sheep graze many winter ranges composed of almost pure stands of big sagebrush (*A. tridentata*) , and the animals receive little forage other than sagebrush for approximately six months of the year. They found big sagebrush to be high in protein, phosphorus, ether extract (fats) , and gross energy. On winter sheep ranges in Utah, big sagebrush produced the highest gross energy value of any forage plant, yielding 5,481 calories of heat per kilogram of plant material (leaves and stems) as compared with 5,010 cal./kg. for alfalfa hay. When compared with other sheep forage plants, the total digestible nutrients and digestible energy for big sagebrush were found to be very high (Table 47). From the standpoint of nutritive content these workers concluded that big sagebrush furnished an adequate diet for pregnant ewes on winter ranges in Utah, and it compared favorably with alfalfa in this respect. Black sagebrush (*A. nova*) was considered by Cook, et al. to be a good forage plant, but, compared with big sagebrush, generally yielded lower digestibility values for most nutrients. They considered big sagebrush to be the least palatable of the species studied, although this conclusion was not necessarily borne out by the amount of dry matter which the sheep consumed daily.

This writer considers palatability of a forage plant to be the result of training and experience of individual animals, and hence difficult to evaluate.

Sagebrush was considered by Smith (1950) to be probably the most important food of mule deer in the intermountain region. The conversion of sagebrush ranges into croplands in many of the intermontane valleys of Wyoming has eliminated thousands of acres of elk and deer winter range and has necessitated costly winter feeding programs. Calf elk-tagging work in Montana disclosed that the majority of calves were dropped in sagebrush types adjacent to timbered lands (Johnson, 1951). Similar findings have been reported from tagging studies on elk calving ranges on the Snake River drainage in Wyoming. The great majority of Wyoming's ninety thousand or more antelope are located in sagebrush-grass ranges, as are the bulk of the antelope herds in the West. The livestock industry of the Mountain West is vitally dependent upon the sagebrush-grass and salt-desert shrub ranges.

In spite of the recommendations and opinions of the various governmental agencies regarding sagebrush, the plant is considered by this author to be the most valuable food and cover plant for both livestock and wildlife on the ranges of the intermountain West. A region which relies so heavily upon its livestock and recreation industries can ill afford to adopt a large-scale sagebrush eradication program founded upon inadequate consideration, ignorance of the basic facts, and the generosities of a government.

TABLE 47

Dry Matter Consumed Daily and Digestibility of Alfalfa and Various Winter Range Plants Found in Foraging Sheep's Diet in Utah
(After Cook et al., 1951)

Forage species	Sheep weight	Weight gain or loss	Dry matter consumed	Digestible dry matter	Digestible energy[*]	Total digestible nutrients in diet	Digestible organic matter	Digestible protein in diet
	Pounds			%		%	%	%
Big sagebrush	151	+0.5	3.28	50.4	3352	74.7	54.8	8.0
Shadscale	150	+0.5	2.93	48.6	1522	33.4	32.3	4.6
Winterfat	147	—1.0	3.16	33.5	1618	35.0	33.8	6.3
Nuttall saltbush	157	+1.5	3.79	26.7	1290	35.5	35.5	2.4
Winterfat and big sagebrush	141	—1.5	3.60	40.1	2273	51.3	41.8	7.5
Winterfat & shadscale	158	0.0	4.50	43.7	1335	31.8	30.6	4.8
Black sagebrush	138	—1.0	3.24	40.7	2315	51.9	42.5	4.4
Squirreltail grass	160	+1.5	4.36	48.7	2021	49.3	48.2	0.0
Alfalfa	140	+1.0	3.65	62.3	2761	57.5	56.2	10.7

[*]Calories per kilogram

Predator Control Activities

Any discussion of predation as related to sage grouse populations must be evaluated in the light of predator control policies now in effect in the western states, and must take into account the forces responsible for the evolution of these policies.

In contrast to the conditions normally encountered by livestock operators in the eastern United States, the western livestock industry has been under actual or potential threat from certain large predators since the beginning of operations. The range cattle business was originally jeopardized by losses to wolves and mountain lions. These predators consequently were exterminated in most states, or at least reduced to levels of impotency. Coyotes, likewise, have constantly represented a hazard to the livestock industry, particularly to range sheep operations. It is entirely possible that large-scale sheep operations have been partially responsible for increasing coyote populations over the western range, especially since sheep numbers rose to incredibly high numbers shortly after 1900. These domestic sheep herds presented easy prey for coyotes since they found them more vulnerable and easier to obtain than wild game and rodents. In the absence of predator control programs, sheep losses from coyotes were excessive.

The Bureau of Biological Survey devoted a large part of its early field activities to lessening predator and rodent pressures on the livestock and ranching industries of the West. The Branch of Predator and Rodent Control, U. S. Fish and Wildlife Service, now functions as the principal agency directing this type of activity. The Branch is supplemented in its control work by the livestock associations, the Bureau of Land Management, the state game departments, and state and local agencies. All of these organizations finance the predator control program on a joint basis, although most of the costs are borne by the livestock operators through the medium of special tax assessments for predator control work.

With the elimination of the wolf and mountain lion from the livestock scene, most of the efforts were directed against the coyote. The control program has been highly effective in reducing sheep losses and in controlling the number of coyotes. Control measures have been intensively applied by highly trained personnel utilizing modern control methods, e.g., 1080 (ten-eighty) and thallium sulphate lethal stations, cyanide guns (coyote-getters), den hunting, snowplanes, and airplanes. The acknowledged control policy of the Fish and Wildlife Service is intensive control of coyotes on livestock ranges. However, this policy is inconsistent with the views of the woolgrowers, the majority of whom necessarily favor extermination of the coyote due to the high cost of continued control and the constant threat of dispersal of coyotes from so-called sanctuary areas. Predator control operations are prohibited in national parks and are greatly restricted on some national forests, particularly in wilderness areas.

As coyote control methods become more intensive and efficient, it can be expected that emphasis will be directed towards control of other mammalian predators such as bobcats and badgers. It may be noted in Table 48 that during the early 1940's, catches of bobcats were relatively high in the four counties of the Green River Basin. During the middle part of the 1940's the catches were considerably reduced. Larger numbers of bobcats were again taken by the hunters during the later years. Prior to 1944 steel traps were the principal means of taking all predatory animals. With the advent of the coyote getters and lethal stations, fewer steel traps were used. Since cyanide guns and lethal stations account for very few bobcats, the catch of this species declined. From 1948 to the present time emphasis has been placed on bobcats, and more steel traps have been put to use by the predator control agents of the Fish and Wildlife Service. It has been of interest to learn recently that in some months the catch of bobcats has exceeded the catch of coyotes in certain counties of southwestern Wyoming.

It has been the opinion of the local agents of the Branch of Predator and Rodent Control that bobcat populations have increased quite widely in the state within recent years. Although bobcats are not considered as serious predators on domestic sheep, the above agency feels that they represent a threat from the standpoint of potential predation and herd disturbance in cases where sheep are grazed on bobcat range. This writer has received many authoritative reports from trappers and livestock operators that bobcats represent little or no hazard to well-managed sheep operations. Accompanying the increase in bobcat populations has been a corresponding increase in badger numbers over range areas. In as much as the normal feeding habits of coyotes, bobcats, and badgers on wild animal prey species are similar in many respects, it is to be expected that any of the predators in this group would increase in numbers as another species in the group was decimated by intensive predator control. Through the application of more intensive control and perhaps a modification of control policies, it would be theoretically possible to exterminate the coyote over the western range, as has been done with the wolf. If any of the large mammalian predators increase, we will likely witness an expanding control program directed against the species on the increase, even though it may not be directly affecting the livestock industry.

Intensive coyote control has not been the only factor contributing to an increase in bobcat and badger populations. In recent years the prices of long-haired furs have been relatively low due chiefly to a high luxury tax and a lack of demand occasioned by style tendencies. There has also been a tendency to blame the decline in value and demand of the long-haired furs upon the importation of foreign furs into the United States, duty free. The recent cancellation of trade concessions to Communist countries may serve to reduce the large quantities of foreign furs being imported into this country. The effect of this embargo may operate to increase the values of domestic furs such as badger, coyote, and bobcat. If the price does not justifiy the efforts of the private trapper in taking these furs, predators may increase in numbers to a point where they will become a nuisance to sage grouse populations.

In support of its predator control policies, the livestock industry has enlisted the aid and support of both sportsmen and state game departments by publicizing and overemphasizing the benefits of predator control in the management of wild game populations.

Predator control programs have been credited with being directly responsible for the high populations of antelope and deer now existing in many western states (Green, 1951). Similar beliefs have been expressed by numerous individuals regarding the increases in sage grouse numbers in recent years. Predator control programs may have been influential in all of these game population increases, but there can be no doubt that the game population increases have been due in greater part to significant changes in the pattern of livestock-use over most western ranges.

In addition to the land improvement programs sponsored by the several land-management agencies, the state and federal governments have been co-operating for more than ten years in a program designed to improve wildlife habitat and game welfare. The activities of this Pittman-Robertson wildlife restoration program likewise have exerted some influence upon game increase.

There is no evidence to indicate that coyote control is directly responsible for the marked increase in numbers exhibited by antelope herds and sage grouse flocks in the center of their ranges. A reduction in predation pressure, or in any other force tending to decimate their numbers, may be expected to cause small groups of antelope and sage grouse to exhibit local increases on submarginal ranges located at the outer limits of their natural distribution. The marked increase in sage grouse (and antelope) numbers in Wyoming has been accompanied by a potentially heavy pressure from coyotes. For example, kill records furnished by the Fish and Wildlife Service have indicated that coyote populations in southwestern Wyoming are maintaining comparatively stable levels of abundance in spite of intensive control methods (Table 48). Since few recoveries are obtained of coyotes killed by lethal stations, it is acknowledged that the annual kill of coyotes is substantially higher than is indicated by the records. This fact is also confirmed by the records of a control agent sponsored by woolgrowers in the Green River Basin. He has stated that the number of coyote pups destroyed annually by him during denning operations has not varied appreciably in the past few years.

Excellent range conditions in Wyoming during the past decade have permitted sage grouse and antelope populations to increase to levels of abundance which, under existing conditions, present a problem in harvesting, at least in some parts of the state. This increase occurred irrespective of losses suffered during the severe winter of 1948-49, a constant threat from coyotes, and in the case of antelope, a hunter harvest approximating 62,000 animals in 1950 and 1951.

The importance of predator control programs in the management of large and widely distributed game populations has been grossly overplayed. Many game officials have been conditioned to transpose the needs and values of predator control operations in the livestock industry to the management of wild game. This reasoning has been characteristic of many western legislative bodies and game departments which normally have close alliance with the livestock industry.

TABLE 48

*Coyote - Bobcat Kill Records for Counties in the Green River Basin from 1941 to 1950**

Year	Lincoln		Uinta		Sublette		Sweetwater		Totals	
	Coyote	Bobcat	C	B	C	B	C	B	C	B
1941	130	12	100	1	52		1507	44	1789	57
1942	407	6	134	11	96		1253	43	1890	60
1943	569		104	1	53		1021	26	1747	27
1944	447	2	58		21		1125	23	1651	25
1945	557	3	273	11	213	1	541	2	1584	17
1946	273	5	220	1	226	1	264	3	983	10
1947	185	1	177	2	210		232	8	804	11
1948	253	13	153	11	209		478	16	1093	40
1949	235	14	258	14	195		483	30	1171	58
1950	137	29	155	25	135		383	21	810	75
Totals	3,193	85	1,632	77	1,410	2	7,287	216	13,522	380

*Based upon data supplied by Branch of Predator and Rodent Control, U. S. Fish and Wildlife Service.

Pursuant to a policy established by the Wyoming Game and Fish Commission in 1946, an annual appropriation of $20,000 has been made available to the U. S. Fish and Wildlife Service as a share in the cost of the statewide predatory animal control program. As part of a public relations program, the Commission has also authorized a predatory bird and small mammal control program in co-operation with the individual counties. Payment for this type of control is established on a 50-50 basis.

Some game administrators still believe that the lack of proper control of predators may be the greatest check to increase of the species in any game habitat; many stress the importance of predatory animal and bird control in the management of game birds and big game. Such beliefs openly expressed at high levels of administration and management make very difficult the task of educating the sporting public to a better understanding of predator-prey relationships in their favorite game habitats.

With few exceptions, predation upon wild game populations has operated not as a force impeding the desirable increase of game, but rather as an important check upon the increase of wildlife numbers above undesired limits. It prevents wild animals from literally taking over an environment at the risk of their own destruction and from reaching levels of abundance threatening conflict with other important land-use activities. Hunting, in the final analysis, is fundamentally a sporting proposition and is limited in its application to prescribed seasonal periods. Under no effective plan of management can hunting function throughout the year either as a decimating agent or as the ultimate force serving to regulate population levels. Hence, other regulatory influences, such as natural predation, are desirable.

PART IV

RESTORATION AND MANAGEMENT

A HISTORY OF SAGE GROUSE RESTORATION

Improvement of Habitat

Restoration of the Western Range

Unrestricted use of the western range has operated in the past as the most serious factor causing the reduction of both livestock and game numbers. During the era of complete exploitation of range resources, certain wildlife populations in the Mountain West decreased almost to the point of extermination.

Overgrazing by domestic livestock is now generally recognized as the factor largely responsible for the decline in numbers of both livestock and wild game during the early years of the twentieth century. Based on the reports of Wyoming's early game protectors, the original decline of sage grouse populations coincided closely with the period of intensive use of range resources by the livestock industry. Poorly planned protective measures, coupled with the drought conditions of the early 1930's, considerably abetted this downward spiral in grouse numbers in the years following the era of maximum range use.

Game animals were first to suffer from deterioration of western ranges. Successive blizzards and a consequent lack of forage on depleted ranges frequently exacted a heavy toll of game. During the period of general game decline (1900-1920), livestock were maintained at higher levels of abundance, since man was able to exercise closer control over domestic herds throughout the seasons. In the years subsequent to 1920, the livestock industry sustained heavy financial losses due to the effects of repeated overuse of the range. This crisis in the livestock industry's economy and the steady decline in game numbers combined to bring about enactment of regulations designed to restore western range lands.

The enactment of the Taylor Grazing Act in 1934 set aside about 132 million acres of western public domain as permanent grazing districts, administered by the Bureau of Land Management and designed to improve western range lands for the sustained production of livestock as well as wildlife. Considerable numbers of sage grouse, antelope, deer, and other wildlife exist on these public lands, and under public ownership they have responded by markedly increasing their numbers. The management practices instituted by the private grazing interests, as well as by the Bureau of Land Management, are affording wildlife a much better place in which to live and reproduce their kind. The excessive overgrazing which was characteristic of past years is gradually being reduced through the combined efforts of the livestock industry and the land-management agencies, both federal and state. This improved program of land use is primarily responsible for the increasing numbers of sage grouse encountered on western range lands during the past decade or so.

Protective Measures

The first regulations providing any sort of protection for sage grouse were promulgated in Wyoming shortly after the turn of the century. These early protective measures were largely concerned with the classification of sage grouse as a legal game species and with the establishment of closed seasons. Several years were to elapse before any limitations were placed on the number of birds which could be killed per day. By the time machinery for the enforcement of necessary protective measures was in operation, the habitat had deteriorated to the point where restoration of sage grouse numbers was more dependent upon a change in the existing land-use policies than on any combination of protective measures. The best laid plans for the protection and increase of most wildlife species, whether they be in the form of legislation or an elaborate and efficient system of law enforcement, are of little avail if the habitat is not physically capable of supporting the species concerned. It is equally axiomatic that the finest wildlife habitat cannot exert its maximum potential in production if handicapped by a lack of protective regulations and personnel qualified to enforce the game laws.

In the case of sage grouse and other typically western game species, an improved game-protective program evolved rather simultaneously with the change in basic policies of range management. The resultant combination materially speeded the recovery of sage grouse populations, particularly since Wyoming, and most other western states, afforded sage grouse complete to partial protection during the period from 1937 to 1948. There is a tendency for some individuals concerned with game management to move rather cautiously before authorizing any measures designed either to increase, or even to inaugurate, the kill of sage grouse in those areas which only a few years ago were so impressively lacking in harvestable numbers of birds. It is well to realize that game survival and increase is not always dependent upon retaining stringent protective regulations, however desirable they may seem to be at the time of their original promulgation.

Replenishment of the Stock

Natural Stocking vs. Trapping and Transplanting

To a large degree, the present increase in sage grouse numbers throughout the West has been accomplished as a result of natural stocking, or dispersal of breeding stock from areas of concentration. Even during the period of low abundance, large numbers of birds undoubtedly thrived in undisturbed areas of their range. The highly migratory nature of the species is believed to have been instrumental in accomplishing the wide dispersion of the species in recent years. The interstate aspects of sage grouse movements have already been demonstrated. Except in a few isolated regions along the periphery of the birds' range, the process of natural stocking would operate effectively in restoring sage grouse populations. If for any reason areas should be depleted of birds, there is relatively little sage grouse habitat which could not be restocked by natural dispersion.

Although the trapping and transplanting of sage grouse in Wyoming during the past ten years has received much credit for the recent increase in numbers, there is little factual data to indicate that this program accomplished anything more than token results. On the basis of band recoveries substantial numbers of adult sage grouse removed from Eden Valley and transplanted into other areas of the Green River Basin returned to the exact locality of their capture within a period of a few weeks. Additional birds returned to Eden Valley after having been released on the Bear River drainage of extreme southwestern Wyoming and on the Wind River drainage of central Wyoming—the birds released on the latter drainage having crossed the southern end of the Wind River Range on their return trip.

After being transplanted into new areas, there was no direct evidence to indicate that immature birds exhibit movements similar to adults. The fate of some several hundred immature sage grouse released simultaneously with the adult birds in 1949 and 1950 remains unknown. In spite of consecutive hunting seasons throughout the region of release in 1950 and 1951, there was no significant return of bands from young birds, either at the sites of capture or release.

Extensive releases of banded birds into Jackson Hole resulted in no marked increase in numbers of resident birds in succeeding years, although strutting ground observations revealed that from 10 to 20 per cent of the transplanted birds actually remained in the vicinity of the release locality and established permanent residence. Most of the birds in this category were known to have been released as juveniles.

The trapping and transplanting program may be of value in attempts to restore depleted populations on a major watershed, *provided* that the habitat is suitable and that a preponderance of young birds are utilized. The seasonal movements of adult birds are so widespread, and the homing instincts are so well-developed that attempts to establish the adult age class on a specific drainage can be expected to result in almost complete failure.

Trapping and banding operations have a definite value in research programs, and they are a distinct aid in studies on grouse longevity, seasonal movements, hunting season kill, etc.

Following their extermination, sage grouse were re-established on a minor scale in New Mexico with stock obtained from Wyoming and South Dakota. Transplanting programs would seem to have definite value in this type of program. There is little merit in promiscuously moving birds into regions which now probably support as many birds as the habitat will accommodate. Sportsmen's clubs are constantly demanding favors of this sort, unaware of the futility of most of these plantings.

Under certain conditions it may be necessary to reduce the number of birds periodically in localized damage areas prior to the hunting season. If such activities are warranted, live-trapping and transplanting can be extremely effective. The birds involved in such a program should be taken at least 50 to 100 miles from the damage area and released in an area which will subsequently be opened to hunting.

A PLAN FOR SAGE GROUSE MANAGEMENT

Control of Range Use

The pattern of range utilization on public lands of the West is now well-established and closely regulated. Livestock and big game grazing allotments on these lands are generally established under joint agreement by the federal land management agencies (i.e., Forest Service, Bureau of Land Management, Indian Service, etc.), the livestock operators (district advisory boards), and the respective state game and fish departments. With certain local exceptions, present range use on public lands in Wyoming is deemed to be consistent with sound range practices and desired carrying capacities; or, if present operations depart from that scheme of management, efforts are being vigorously directed toward the accomplishment of that objective.

The wide range of landownership types necessary for year-round livestock operations dictates to a large degree that the pattern of use on the private ranges of the West will conform essentially to the intensity of use which is practiced on the public range lands.

Use of the public and private range lands of the upper Green River Basin by livestock (primarily sheep, cattle, and horses) and wild game (mainly sage grouse, antelope, and mule deer) during the past fifteen years has resulted in no deterioration of range lands. One the other hand, the numbers of game and domestic animals (notably cattle) ranging in the Basin have steadily increased during the period specified, further indicating that the present land-control policies are a distinct boon to livestock as well as wildlife interests. Several hundred thousand sheep and cattle share seasonal ranges with a like number of sage grouse. Several thousand antelope and mule deer, and lesser numbers of elk and moose, likewise thrive on private and public range lands in the Basin. Innumerable areas may be found in Wyoming and other western states where range carrying capacities have been increased under the existing system of control (whereby land-use practices and allotments emanate jointly from national, state, and private levels of administration). Sage grouse are primarily browsing animals, and as such are indirectly affected by the management practices conducted for the benefit of hoofed mammals.

Reclamation Programs

The programs of reclaiming sagebrush lands for agricultural development, regardless of their merit, pose the most serious threat to the future welfare of sage grouse populations.

The complete restoration of the original extent of sage lands in the Mountain West will never be possible. Agricultural and power developments, vital as they are to our economy, must inevitably destroy a portion of the primitive vegetative types, as well as the animals which they support. As a result of these developments, some of the finest sagebrush lands and sage grouse habitat in the West have already been destroyed. This destruction of habitat is continuing at an accelerated pace under the pressure of an expanding reclamation program.

The states must direct greater efforts and more study to the long range values of reclamation programs in their respective domains before extending blanket approval for the completion of myriads of power and water development projects. Interstate compacts regulate water rights and water use on every major drainage in western United States. The state of Wyoming is allocated 14 per cent of the average annual flow of the upper Colorado River Basin for industrial and agricultural development.

The use of water by wildlife or recreational users is established purely on a by-product basis; seldom is any allocation of water made to these interests which in recent years have loomed as primary sources of revenue in Wyoming, as well as in certain other western states.

The welfare of wildlife is so affected by reclamation projects that definite regulations should be adopted concerning the conflicts which inevitably arise between the sponsoring agencies, the private landowners or homesteaders, and the state game and fish departments.

Regardless of the game species involved, just consideration should be given to a program designed to compensate the state for reductions in habitat available to particular game animals as a result of the project plans. This could either take the form of land exchanges, which could ultimately be developed as a means of offsetting the undesirable features of the reclamation program as they affect wildlife, or else through a cash settlement to be used by the state in accomplishing the same goal.

Additional consideration should be given to the formulation of a plan to compensate the future homesteader for damage which will be inflicted at some future date upon his private holdings as a result of thrusting agricultural developments into the midst of extensive game range. It is rather inconsistent to expect or demand that the sportsmen of any state must stand liable for damage payments of this nature. This compensation could take any number of forms, but should be incorporated in the original contract between the landowner and the government. The payment of damage claims, in one form or another, is placing increasingly heavy demands upon western game and fish departments. Under the present method of deriving game and fish funds (license and permit sales), damage payments may reach limits intolerable to most game and fish departments. In the absence of any other methods for taking care of damage, departments are forced to reduce drastically, and in some cases completely eliminate, game populations which in all other respects are a desirable feature of the states' economy.

As previously stated, a complete evaluation of the effects of the project upon wildlife welfare is not always forthcoming. either on the part of the state or federal government. In the case of sage grouse, year-round investigations must be conducted in order to properly evaluate the seasonal movements and the extent of range necessary for their survival.

Full co-operation between federal, state, and private agencies is yet to be attained in resolving wildlife problems encountered on river basin developments. Until that goal is reached, wildlife interests will continue to suffer indignities at the hands of the control agencies.

Control of Harvest

Restricting sage chicken hunting to the immediate vicinity of ranching areas in Eden Valley will be ineffective as a method of alleviating future damage to alfalfa crops. A large percentage of the adult birds (cocks in particular) retire to outlying areas while the hunting season is in progress. The seasonal movements of sage grouse in the Green River Basin would likely restore the Eden Valley populations to peak levels regardless of the intensity of hunting and the extent of the kill in that limited area. The relatively low hunting pressures exerted on sage grouse during the 1950 and 1951 hunting seasons indicated that more liberal regulations may be necessary in order to encourage hunters from other areas of the state to participate in the hunt. It may be necessary to lengthen the seasons, open larger areas to hunting, and modify bag and possession limits in succeeding years in an effort to attract more hunters. In view of the high populations of sage grouse now existing in the Green River Basin, as well as in certain other areas of the state, it would seem desirable to establish regulations designed to increase both the hunting pressure and the kill. During years of so-called cyclic low grouse populations, a modification in regulations may be necessary. If past trends in grouse numbers are repeated in the future, we can expect years of high grouse abundance to be interspersed with years of low grouse abundance, due mainly to causes presently beyond our control.

Hunting regulations must afford sage grouse maximum protection during the brooding period and throughout the dry summer season. In areas permitting an annual harvest, hunting seasons should be delayed until early fall. At this time of the year sage grouse are well-distributed throughout their habitat and are not concentrated around desert oases. Bag limits and season lengths should be governed by the numbers of birds available to be harvested and corresponding hunting pressures, rather than on standardized regulations approved in a perfunctory manner each year.

State game and fish departments must further recognize the desirability of harvesting surplus numbers of sage grouse, along with other game, in those areas supporting high numbers of birds and during the exact years in which those surpluses exist. In few western states, will either present or future populations of sage grouse permit hunting seasons on a wide regional or state-wide basis. Seven western states allowed sage grouse hunting in 1950 and 1951, either on a special permit or on a generalized open season basis, indicating the present trend in management of this bird. It is believed that Colorado and Montana could also harvest sage grouse in certain areas under systems of hunting appropriate to their state.

Game enforcement officials in southwestern Wyoming were impressed with the noticeable reduction in the incidence of sage grouse poaching in their districts in 1950 and 1951. Wardens believed that the tendency for hunters to show increasing respect for the game bird law was a direct outgrowth of opening areas of grouse concentration to legal hunting in those years.

It is again emphasized that sage grouse hunting seasons in the mountainous states should not be conducted prior to September 20.

Establishment of Permanent Census Areas

In contrast to the situation which existed a few years ago, most western game and fish departments have assigned biologists to work on permanent sage grouse projects. Other departments contemplate setting up similar projects as an aid to management. The importance of sage grouse in a state game bird program warrants continuous study of population trends and seasonal movements, and a constant appraisal of land-use programs threatening to destroy the birds' habitat.

The relative ease with which yearly population trends can be determined justifies the establishment of permanent census tracts on the major watersheds which furnish annual sport from sage grouse hunting. Efforts should be directed toward strutting ground and roadside brood censuses due to the simplicity of making these counts and the reliability of the data so obtained.

Survival Outlook for Sage Grouse Populations Under Land-Use Programs of the Future

As a result of the factors previously outlined, sage grouse have been restored in many western areas to levels of abundance characteristic of the years of plenty.

In the semiarid regions of the intermountain area, landowner-ship and land-utilization patterns are constantly changing, particularly on those localized areas which may be accessible to water. The development of these lands for other land-use activities will jeopardize sage grouse populations, either through the elimination of breeding areas or by a reduction of wintering ranges. The fate of sage grouse, as well as antelope and other associated wildlife species, will be dependent upon the degree of maintenance and preservation afforded the vast tracts of sage lands in the West.

Sage grouse are somewhat fortunate in being able to share ranges which, in addition to furnishing a livelihood for game, provide the very sustenance necessary to the survival of the western livestock industry. An economy as well-established as grazing is not likely to be displaced from the western scene. Neither are the sagebrush lands dedicated to grazing likely to disappear from the livestock picture, since stockmen almost unanimously favor their preservation and mantenance. A continuation of these attitudes and the present grazing policies will go far toward assuring the permanency of sage grouse populations.

Sportsmen and ornithologists, who for years have been pondering the fate of sage grouse, may now rest assured that the species is in no immediate danger of extinction. Certain land-use programs will periodically decimate localized sage grouse populations. In the absence of any better control, these programs will ultimately be limited in their scope by the amount and availability of water supplies on the semiarid lands of the West. Once that point is reached, sagebrush lands should be spared further exploitation; and only then will we know what future degree of security lies in store for the sage grouse.

PART V

APPENDIX

TABLE 49

Trapping and Transplanting Sites of Sage Grouse—1949

Site of Capture (Ranch)		Date of Release	No. of Birds	Site of Release	County
McComas	—Eden	Aug. 18-49	31	Antelope Flat	Teton
McComas	—Eden	Aug. 20-49	25	Antelope Flat	Teton
McComas	—Eden	Aug. 24-49	17	Antelope Flat	Teton
Carlson	—Eden	Aug. 26-49	120	Fed.ElkRefuge	Teton
Carlson	—Eden	Aug. 31-49	21	Antelope Flat	Teton
McMurray	—Eden	Sept. 3-49	22	Fed.ElkRefuge	Teton
McMurray	—Eden	Sept. 3-49	71	Blacktail Butte	Teton
McMurray	—Eden	Sept. 5-49	78	Bear River Div.	Uinta
Nelson	—Eden	Sept. 11-49	66	Bear River Div.	Uinta
Arenbelle	—Elkhorn	Sept. 13-49	89	Bear River Div.	Uinta
Arenbelle	—Elkhorn	Sept. 13-49	61	Bear River Div.	Uinta
McComas	—Eden	Sept. 17-49	88	Alma	Uinta
Chesnovar	—Farson	Sept. 23-49	19	Alma	Uinta
McComas	—Eden	Sept. 23-49	72	Alma	Uinta
McComas	—Eden	Sept. 26-49	5	Alma	Uinta
McMurray	—Eden	Sept. 26-49	19	Alma	Uinta
Swanstrom	—Farson	Sept. 26-49	17	Alma	Uinta
Carlson	—Eden	Sept. 27-49	113	Ring Mt.	Albany
Gregg	—Eden	Oct. 1-49	92	Hoback Rim	Sublette
McMurray	—Eden	Oct. 7-49	49	Eden Valley	Sw'twat'r
Gregg	—Eden	Oct. 13-49	8	Eden Valley	Sw'twat'r
Gregg	—Eden	Oct. 15-49	9	Eden Valley	Sw'twat'r
Swanstrom	—Farson	Oct. 17-49	7	Eden Valley	Sw'twat'r
Total			1,099		

TABLE 50

Trapping and Transplanting Sites of Sage Grouse—1950

Site of Capture (Ranch)	Date of Release	No. of Birds	Site of Release	County
McComas —Eden	July 8-50	9	Sublette Springs	Sublette
McComas —Eden	July 9-50	26	Gaston Crossing	Sweetwater
Dearth —Eden	July 13-50	15	Eden Valley	Sweetwater
Carlson —Eden	July 14-50	7	Eden Valley	Sweetwater
Carlson —Eden	July 14-50	37	Superior Oil Well	Sublette
Joslin —Eden	July 18-50	25	Eden Valley	Sweetwater
Joslin —Eden	July 18-50	8	Eden Valley	Sweetwater
Dearth —Eden	July 19-50	69	Eden Valley	Sweetwater
Dearth —Eden	July 19-50	57	Big Island Bridge	Sweetwater
Dearth —Eden	July 20-50	57	14 Mile Hill	Sweetwater
Dearth —Eden	July 21-50	39	15 Mile Spring	Sweetwater
Joslin —Eden	July 22-50	11	15 Mile Spring	Sweetwater
Nelson —Eden	July 23-50	46	Hays Middle Rch.	Sweetwater
McMurray—Eden	July 30-50	89	Elkhorn	Sublette
McMurray—Eden	July 31-50	40	Sand Springs	Sublette
McComas —Eden	Aug. 1-50	67	Chalk Butte	Sublette
McComas —Eden	Aug. 3-50	70	Five Mile Creek	Sheridan
Nelson —Eden	Aug. 3-50	104	Five Mile Creek	Sheridan
Murphy —Eden	Aug. 9-50	125	Birmingham Rch.	Fremont
Murphy —Eden	Aug. 8-50	5	Eden Valley	Sweetwater
Gregg —Eden	Aug. 11-50	125	Five Mile Creek	Sheridan
McComas —Eden	Aug. 14-50	92	N.W. of Midwest	Johnson
McComas —Eden	Aug. 14-50	89	N.W. of Midwest	Johnson
Joslin —Eden	Aug. 17-50	55	Birmingham Rch.	Fremont
Gregg —Eden	Aug. 17-50	46	Birmingham Rch.	Fremont
Gregg —Eden	Aug. 19-50	7	Eden Valley	Sweetwater
Grandy —Farson	Aug. 19-50	99	Birmingham Rch.	Fremont
McMurray—Eden	Aug. 26-50	31	Gierce Ranch	Johnson
Laughlin —Eden	Aug. 26-50	96	Gierce Ranch	Johnson
McComas —Eden	Aug. 26-50	82	Gierce Ranch	Johnson
Dearth —Eden	Aug. 26-50	22	Eden Valley	Sweetwater
Dearth —Eden	Aug. 26-50	27	Eden Valley	Sweetwater
Nelson —Eden	Oct. 7-50	93	Arkansas Creek	Sheridan
Nelson —Eden	Oct. 7-50	70	Arkansas Creek	Sheridan
Rahm —Farson	Oct. 12-50	19	S.E. Rock Springs	Sweetwater
Rahm —Farson	Oct. 12-50	12	Eden Valley	Sweetwater
Nelson —Eden	Oct. 15-50	47	New So. Pass City	Fremont
Rahm —Farson	Oct. 19-50	4	Eden Valley	Sweetwater
Nelson —Eden	Oct. 20-50	19	Eden Valley	Sweetwater
Stout —Eden	Oct. 21-50	7	Eden Valley	Sweetwater
Nelson —Eden	Oct. 23-50	9	14 Mile Hill	Sweetwater
Skorz —Eden	Oct. 27-50	29	Red Canyon	Fremont
Skorz —Eden	Oct. 27-50	4	Red Canyon	Fremont
Strutting Grounds	May 9-50	20	Eden Valley	Sweetwater
Total		2,010		

TABLE 51

*A Partial Check List of the Mammals Occurring with Sage Grouse on the Sagebrush Plains of the Upper Green River Basin**

CLASS *MAMMALIA*—mammals

Order *INSECTIVORA*—moles and shrews

Family *Soricidae*—shrews

Sorex cinereus cinereus	cinereus shrew
Sorex merriami merriami	Merriam shrew
Sorex vagrans monticola	wandering shrew

Order *CHIROPTERA*—bats

Family *Vespertilionidae*—vespertilionid bats

Myotis lucifugus carissima	big myotis
Myotis evotis evotis	long-eared myotis
Myotis volans interior	hairy-winged myotis
Myotis subulatus subulatus	small-footed myotis
Lasionycteris noctivagans	silver-haired bat
Corynorhinus rafinesquei pallescens	long-eared bat
Eptesicus fuscus pallidus	big brown bat
Lasiurus cinereus cinereus	hoary bat

Order *CARNIVORA*—flesh-eating mammals

Family *Mustelidae*—weasels and allies

Mustela frenata nevadensis	long-tailed weasel
Mustela vison energumenos	mink
Spilogale gracilis saxatilis	Great Basin spotted skunk
Mephitis mephitis hudsonica	striped skunk
Taxidea taxus montana	badger

Family *Canidae*—dogs and foxes

Canis latrans lestes	coyote

Family *Felidae*—cats

Felis concolor hippolestes	mountain lion
Lynx rufus pallescens	bobcat

Order *RODENTIA*—gnawing mammals

Family *Sciuridae*—squirrels and allies

Eutamias minimus minimus	least chipmunk
Citellus richardsoni elegans	Wyoming ground squirrel
Citellus armatus	Uinta ground squirrel
Citellus tridecemlineatus parvus	thirteen-lined ground squirrel
Cynomys leucurus	white-tailed prairie dog

Family *Castoridae*— beavers
   Castor canadensis duchesnei   beaver
Family *Heteromyidae*—pocket mice and kangaroo rats
   Perognathus parvus clarus   Great Basin pocket mouse
   Dipodomys ordi priscus   kangaroo rat
Family *Geomyidae*—pocket gophers
   Thomomys talpoides ocius   northern pocket gopher
Family *Muridae*—Old World rats and mice
   Mus musculus musculus   house mouse
Family *Cricetidae*—hamster-like rats and mice
   Peromyscus maniculatus osgoodi   deer mouse
   Onychomys leucogaster arcticeps northern grasshopper mouse
   Neotoma cinerea orolestes   bushy-tailed woodrat
   Microtus montanus caryi   Montana meadow vole
   Microtus pennsylvanicus modestus Pennsylvania meadow vole
   Lagurus curtatus levidensis   sagebrush vole
   Ondatra zibethica osoyoosensis   muskrat
Family *Erethizontidae*—American porcupines
   Erethizon dorsatum epixanthum   porcupine

Order *LAGOMORPHA*—rabbits, pikas, and hares

Family *Leporidae*—hares and rabbits
   Lepus townsendi campanius   white-tailed jack rabbit
   Sylvilagus nuttalli grangeri   Nuttall cottontail
   Sylvilagus auduboni baileyi   Audubon cottontail

Order *ARTIODACTYLA*—even-toed hoofed mammals

Family *Cervidae*—deer and allies
   Cervus canadensis nelsoni   Yellowstone elk
   Odocoileus virginianus macrurus western white-tailed deer
   Odocoileus hemionus hemionus   Rocky Mountain mule deer
   Alces americana shirasi   Wyoming moose
Family *Antilocapridae*—pronghorn
   Antilocapra americana americana American pronghorn

*List includes forms observed by the writer in sagebrush types of
the Upper Green River Basin, and supplemented with the published and unpublished records of mammalogists.

TABLE 52

A Partial Check List of the Birds Occurring with Sage Grouse on the Sagebrush Plains of the Upper Green River Basin *

CLASS *AVES*—birds

Order *COLYMBIFORMES*—grebes

Family *Colymbidae*—grebes

Colymbus auritus	horned grebe—M
Aechmophorus occidentalis	western grebe—R
Podilymbus podiceps	pied-billed grebe—R

Order *PELECANIFORMES*—fully-webbed swimmers

Family *Pelecanidae*—pelicans

Pelecanus erythrorhynchos	white pelican—M

Order *CICONIIFORMES*—deep-water waders

Family *Ardeidae*—herons and bitterns

Ardea herodias	great blue heron—B
Leucophoyx thula	snowy egret—M
Botaurus lentiginosus	American bittern—M

Family *Threskiornithidae*—ibises

Plegadis mexicana	white-faced glossy ibis—M

Order *ANSERIFORMES*—sieve-billed swimmers

Subfamily *Cygninae*—swans

Cygnus columbianus	whistling swan—M

Subfamily *Anserinae*—geese

Branta canadensis	Canada goose—B
Anser albifrons	white-fronted goose—M
Chen hyperborea	lesser snow goose—M

Subfamily *Anatinae*—surface-feeding ducks

Anas platyrhynchos	mallard—B
Anas strepera	gadwall—B
Mareca americana	baldpate—B
Anas acuta	pintail—B
Anas carolinensis	green-winged teal—B
Anas discors	blue-winged teal—B
Anas cyanoptera	cinnamon teal—B
Spatula clypeata	shoveller—B

Subfamily *Nyrocinae*—diving ducks
  Aythya americana   redhead—M
  Aythya collaris   ring-necked duck—M
  Aythya valisineria   canvas-back—M
  Aythya affinis   lesser scaup duck—M
  Bucephala clangula   American golden-eye—WR
  Bucephala islandica   Barrow's golden-eye—WR
  Bucephala albeola   buffle-head—M
Subfamily *Erismaturinae*—ruddy and masked ducks
  Oxyura jamaicensis   ruddy duck—M
Subfamily *Merginae*—mergansers
  Lophodytes cucullatus   hooded merganser—B
  Mergus merganser   American merganser—B
  Mergus serrator   red-breasted merganser—M

Order *FALCONIFORMES*—diurnal birds of prey

Subfamily *Accipitrinae*—accipiters, or short-winged hawks
  Accipiter gentilis   goshawk—WR
  Accipiter striatus   sharp-shinned hawk—B
  Accipiter cooperii   Cooper's hawk—B
Subfamily *Buteoninae*—buteos, or buzzard hawks, and eagles
  Buteo jamaicensis   red-tailed hawk—B
  Buteo swainsoni   Swainson's hawk—B
  Buteo lagopus   American rough-legged hawk—WR
  Buteo regalis   ferruginous rough-legged hawk—B
  Aquila chrysaetos   golden eagle—B
  Haliaetus leucocephalus bald eagle—M
Subfamily *Circinae*—harriers
  Circus cyaneus   marsh hawk—B
Subfamily *Pandioninae*—ospreys
  Pandion haliaetus   osprey—M
Subfamily *Falconinae*—falcons
  Falco mexicanus   prairie falcon—B
  Falco peregrinus   duck hawk—M
  Falco columbarius   pigeon hawk—B
  Falco sparverius   sparrow hawk—B

Order *GALLIFORMES*—gallinaceous birds

Family *Tetraonidae*—grouse
  Centrocercus urophasianus sage grouse—B
Family *Phasianidae*—pheasants
  Phasianus colchicus   ring-necked pheasant—B

Order *GRUIFORMES*—cranes and rails

Family *Gruidae*—cranes
 Grus canadensis   sandhill crane—M

Family *Rallidae*—rails, coots, and gallinules
 Porzana carolina   sora—R
 Fulica americana   coot—B

Order *CHARADRIIFORMES*—shore birds and gulls

Family *Charadriidae*—plovers
 Eupoda montana   mountain plover—B
 Charadrius vociferus   killdeer—B

Family *Scolopacidae*—snipe, sandpipers, etc.
 Capella gallinago   Wilson's snipe—R
 Numenius americanus   long-billed curlew—R
 Actitis macularia   spotted sandpiper—R
 Catoptrophorus semipalmatus western willet—M
 Totanus melanoleucus   greater yellow-legs—M
 Totanus flavipes   lesser yellow-legs—M
 Limnodromus griseus   long-billed dowitcher—M
 Limosa fedora   marbled godwit—M

Family *Recurvirostridae*—avocets and stilts
 Recurvirostra americana   avocet—B

Family *Phalaropodidae*—phalaropes
 Steganopus tricolor   Wilson's phalarope—B
 Lobipes lobatus   northern phalarope—M

Family *Laridae*—gulls and terns
 Larus californicus   California gull—M
 Larus pipixcan   Franklin's gull—M

Order *COLUMBIFORMES*—pigeons and doves

Family *Columbidae*—pigeons and doves
 Columba livia   domestic pigeon—B
 Zenaidura macroura   mourning dove—B

Order *STRIGIFORMES*—nocturnal birds of prey

Family *Strigidae*—typical owls
 Bubo virginianus   horned owl—B
 Speotyto cunicularia   burrowing owl—B
 Asio otus   long-eared owl—R
 Asio flammeus   short-eared owl—B

Order *CAPRIMULGIFORMES*—goatsuckers

Family *Caprimulgidae*—goatsuckers
  Chordeiles minor   nighthawk—B

Order *CORACIIFORMES*—kingfishers

Family *Alcedinidae*—kingfishers
  Megaceryle alcyon   belted kingfisher—R

Order *PICIFORMES*—woodpeckers

Family *Picidae*—woodpeckers
  Colaptes cafer   red-shafted flicker—R
  Asyndesmus lewis   Lewis's woodpecker—R
  Sphyrapicus varius   red-naped sapsucker—R
  Sphyrapicus thyroideus   Williamson's sapsucker—R
  Dendrocopos pubescens   downy woodpecker—R

Order *PASSERIFORMES*—perching birds

Family *Tyrannidae*—flycatchers
  Tyrannus tyrannus   eastern kingbird—R
  Tyrannus verticalis   western kingbird—R

Family *Alaudidae*—larks
  Eremophila alpestris   horned lark—B

Family *Hirundinidae*—swallows
  Iridoprocne bicolor   tree swallow—B
  Stelgidopteryx ruficollis   rough-winged swallow—B
  Hirundo rustica   barn swallow—B
  Petrochelidon pyrrhonata   cliff swallow—B

Family *Corvidae*—crows and jays
  Pica pica   American magpie—B
  Corvus corax   raven—R
  Corvus brachyrhynchos   crow—M
  Nucifraga columbiana   Clark's nutcracker—R

Family *Paridae*—titmice
  Parus atricapillus   black-capped chickadee—R
  Parus gambeli   mountain chickadee—R

Family *Sittidae*—nuthatches
  Sitta carolinensis   white-breasted nuthatch—R
  Sitta canadensis   red-breasted nuthatch—R

Family *Cinclidae*—dippers or water ouzels
  Cinclus mexicanus   dipper—R

Family *Troglodytidae*—wrens
 Troglodytes aedon   house wren—B
 Salpinctes obsoletus   rock wren—B

Family *Mimidae*—mocking birds and thrashers
 Mimus polyglottos   mockingbird—B
 Dumetella carolinensis   catbird—R
 Oreoscoptes montanus   sage thrasher—B

Family *Turdidae*—robins, bluebirds, and thrushes
 Turdus migratorius   robin—B
 Sialia currucoides   mountain bluebird—B

Family *Sylviidae*—kinglets
 Regulus calendula   ruby-crowned kinglet—M

Family *Bombycillidae*—waxwings
 Bombycilla cedrorum   cedar waxwing—M

Family *Laniidae*—shrikes
 Lanius ludovicianus   loggerhead shrike—B

Family *Sturnidae*—true starlings
 Sturnus vulgaris   starling—WR

Family *Vireonidae*—vireos
 Vireo gilvus   warbling vireo—R

Family *Compsothlypidae*—wood warblers
 Dendroica petechia   yellow warbler—B
 Dendroica auduboni   Audubon's warbler—R
 Oporornis tolmiei   Macgillivray's warbler—R
 Wilsonia pusilla   pileolated warbler—M

Family *Ploceidae*—weaver finches
 Passer domesticus   English sparrow—B

Family *Icteridae*—blackbirds, orioles, and meadowlarks
 Dolichonyx oryzivorous   bobolink—R
 Sturna neglecta   western meadowlark—B
 Xanthocephalus xanthocephalus yellow-headed blackbird—B
 Agelaius phoeniceus   common redwing—B
 Icterus bullockii   Bullock's oriole—B
 Euphagus cyanocephalus   Brewer's blackbird—B

Family *Thraupidae*—tanagers
 Piranga ludociviana   western tanager—M

Family *Fringillidae*—sparrows, finches, grosbeaks, and buntings

Pheucticus melanocephalus	black-headed grosbeak—R
Leucosticte tephrocotis	gray-crowned rosy finch—R
Spinus pinus	pine siskin—R
Spinus tristis	common goldfinch—R
Chlorura chlorura	green-tailed towhee—B
Calamospiza melanocorys	lark bunting—B
Passerculus sandwichensis	savannah sparrow—B
Pooecetes gramineus	vesper sparrow—B
Chondestes grammacus	lark sparrow—R
Amphispiza belli	sage sparrow—B
Junco oreganus	pink-sided junco—R
Spizella passerina	chipping sparrow—B
Spizella breweri	Brewer's sparrow—B
Zonotrichia leucophrys	white-crowned sparrow—R
Melospiza melodia	song sparrow—B
Plectrophenax nivalis	snow bunting—WR

*Species observed by the writer in sagebrush types of the upper Green River Basin during the period from May, 1948 to November, 1951.

B Known to breed in the Basin, though not necessarily in sagebrush types.

M Regarded as a spring and fall migrant in the Basin.

R Summer resident and assumed to breed in the Basin.

WR Winter resident in the Basin.

TABLE 53

The Groups of Foods Most Commonly Eaten by Adult Sage Grouse in Wyoming During Each Season

PLANTS

Spring*		Summer*	
Species	Volume per cent	Species	Volume per cent
Sagebrush	86.5	Sagebrush	44.9
Sweet clover	10.7	Alfalfa	13.6
Rabbitbrush	1.0	Dandelion	11.2
Dandelion	.8	Rabbitbrush	5.7
Grasses	.02	Vetches	4.8
		Salsify	4.3

Fall*		Winter*	
Sagebrush	80.7	Sagebrush	99.7
Dandelion	6.6		
Alfalfa	5.1		
Clover	3.2		
Prickly lettuce	2.6		
Knotweed	.8		

ANIMALS

Spring*		Summer*	
Leaf beetles (*Chrysamelidae*)	.2	Grasshoppers (*Locustidae*)	9.4
Ladybird beetles (*Coccinellidae*)	.2	Insect galls	1.6
Insect galls	.1	Ants (*Formicidae*)	.6
Ants (*Formicidae*)	.1	Sand cricket (*Stenopelmatus*)	.3
Darkling beetles (*Tenebrionidae*)	.1	Leaf beetles	.2
		Ladybird beetles	.1
		Darkling beetles	.1

Fall*		Winter*	
Grasshoppers	4.2	Insect galls	.3
Carab beetles (*Carabidae*)	.4		
Ladybird beetles	.3		
Ants	.1		

*Spring—April and May
*Summer—June, July, and August
*Fall—September and October
*Winter—November, December, January, February, and March

TABLE 54

The Groups of Foods Most Commonly Eaten by Juvenile Sage Grouse in Wyoming During the Spring and Summer

PLANTS

Spring		Summer	
Species	Volume per cent	Species	Volume per cent
Clover (*Trifolium*)	13.5	Sagebrush	46.7
Buckwheat (*Polygonum*)	3.6	Rabbitbrush	
Sagebrush (*Artemisia*)	3.5	(*Chrysothamnus*)	12.9
Dandelion (*Taraxacum*)	3.2	Dandelion	6.1
		Sweet clover (*Melilotus*)	5.8
		Prickly lettuce (*Lactuca*)	4.8
		Grasses	3.3
		Alfalfa (*Medicago*)	2.8
		Peppergrass (*Lepidium*)	1.7
		Salsify (*Tragopogon*)	1.2

ANIMALS

Species	Volume per cent	Species	Volume per cent
Ants (*Formica rufa* and Tapinoma sessile*)	27.6	Ants (*Formicidae*)	5.4
Beetles (*Aphodius*)	22.5	Grasshoppers (*Melanoplus*)	3.1
Weevils (*Curculionidae*)	12.1	Insect galls	.8
Beetles (*Carabidae*)	2.7	Ladybird beetles	
		(*Coccinellidae*)	.7
Flies (*Diptera*)	.9	Leaf beetles	
		(*Chrysamelidae*)	.7
Moths (*Lepidoptera*)	.9	Darkling beetles	
		(*Tenebrionidae*)	.3
		Weevils (*Curculionidae*)	.1
		Larvae	.1

LITERATURE CITED

Allen, Ena A. 1934. *Eimeria angusta* n. sp. and *Eimeria bonasae* n. sp. from grouse, with a key to the species of *Eimeria* in birds. Trans. Amer. Micros. Soc. LIII (1) : 1-5.

Allred, Warren J. 1950. Re-establishment of seasonal elk migration through transplanting. Trans. N. Amer. Wildl. Conf. 15 : 597-611.

Bailey, Alfred M. 1925. Segregation of the sexes in the sage-hen. Condor 27 (4) : 172-173.

Batterson, Wesley M., and Wm. B. Morse. 1948. Oregon sage grouse. Ore. Game Comm. Fauna Series 1 : 1-29.

Beath, O. A., C. S. Gilbert, and H. F. Eppson. 1939. The use of indicator plants in locating seleniferous areas in western United States. Amer. Jour. Botany 26 (4) : 257-269.

Bent, A. C. 1932. Life histories of North American gallinaceous birds. Orders *Galliformes* and *Columbiformes*. U. S. Nat. Mus. Bull. 162 : 300-310.

Bergeson, Wm. R. 1951. Game bird biologist, Mont. Dept. Fish & Game. Personal letter, Jan. 22, 1951.

Bonaparte, Charles Lucian. 1827. Notice of a nondescript species of grouse. Zool. Jour. III : 212-213.

Bond, Frank. 1900. A nuptial performance of the sage cock. Auk. 17 (4) : 325-327.

Brooks, Allan C. 1930. The specialized feathers of the sage hen. Condor 32 (4) : 205-207.

Brues, Charles T. 1946. Insect dietary. An account of the food habits of insects. Harvard University Press : 1-466.

Bump, Gardiner, Robert W. Darrow, Frank C. Edminster, and Walter F. Crissey. 1947. The ruffed grouse. New York State Cons. Dept. : 1-915.

Burnett, L. E. 1905. The sage grouse, *Centrocercus urophasianus*. Condor 7 (4) : 102-105.

Buss, Irven O., Roland K. Meyer, and Cyril Kabat. 1951. Wisconsin pheasant reproduction studies based on ovulated follicle technique. Jour. Wildl. Mgt. 15 (1) : 32-46.

Clarke, L. Floyd, Hermann Rahn, and Malcolm D. Martin. 1942. Sage grouse studies. Part II. Seasonal and sexual dimorphic variations in the so-called "air sacs" region of the sage grouse. Wyo. Game & Fish Dept. Bull. No. 2 : 13-27.

Cook, C. Wayne, and Lorin E. Harris. 1950. The nutritive content of the grazing sheep's diet on the summer and winter ranges of Utah. Utah State Agric. Coll. Expt. Sta. Bull. 342 : 1-66.

Cook, C. Wayne, L. A. Stoddart, and Lorin E. Harris. 1951. Measuring consumption and digestibility of winter range plants by sheep. Jour. Range Mgt. 4 (5) : 335-346.

Cottam, Clarence. 1930. In charge of the Section of Food Habits Research, U. S. Biological Survey. Personal letter to George L. Girard, March 26, 1930.

Cottam, Walter P., and George Stewart. 1940. Plant succession as a result of grazing and of meadow dessiccation by erosion since settlement in 1862. Jour. Forestry 38 (8) : 613-626.

Coues, Elliott. 1874. Birds of the Northwest. A handbook of the ornithology of the region drained by the Missouri River and its tributaries. U. S. Dept. Int., U. S. Geol. Surv. of Territories, Misc. Publ. No. 3 : 400-407.

Coulter, John M., and Aven Nelson. 1909. New manual of botany of the central Rocky Mountains. American Book Co., New York : 1-646.

Craighead, F. C., Jr., and John J. Craighead. 1949. Nesting Canada geese on the upper Snake River. Jour. Wildl. Mgt. 13 (1) : 51-64.
                                          1950. Ecology of raptor predation. Unpubl. Ph.D. dissertations, Univ. of Michigan Library, Ann Arbor.

Fremont, John C. 1845. Report of the exploring expedition to the Rocky Mountains in the year 1842, and to Oregon and North California in the years 1843-44. Blair and Rives, Printers, Washington.

Girard, George L. 1937. Life history, habits, and food of the sage grouse, *Centrocercus urophasianus* Bonaparte. Univ. of Wyo. Publications 3 (1) : 1-56.

Grange, Wallace B. 1949. The way to game abundance, with an explanation of game cycles. Charles Scribner's Sons, New York : 1-365.

Green, Dorr D. 1951. Modern methods of predator control. Trans. Intern. Assoc. of Game, Fish, and Conservation Commissioners 41 : 114-121.

Griner, Lynn A. 1939. A study of the sage grouse (*Centrocercus urophasianus*), with special reference to life history, habitat requirements, and numbers and distribution. Unpubl. Master's thesis, Utah State Agric. Coll. Library, Logan.

Grinnell, Joseph, H. C. Bryant, and T. I. Storer. 1918. The game birds of California. Univ. of Calif. Semi-Centennial Publ. : 564-572.

Hall, Harvey M., and Frederic E. Clements. 1923. The North American species of *Artemisia*, *Chrysothamnus* and *Atriplex*. Carnegie Institution of Wash. Publ. No. 326.

Hamerstrom, F. N., Jr. 1940. Wisconsin prairie grouse studies, 1929-40. Unpubl. Ph.D. dissertation, Univ. of Wisc. Library, Madison.

Henderson, Harry. 1951. State Game Warden, Butte County, South Dakota. Personal letter, Jan. 29, 1951.

Holscher, Clark E. 1944. Controlling of prickly pear. Western Livestock Jour. 22 (42).

Honess, Ralph F. 1941. Sage grouse wintering in Jackson Hole, Wyoming. Unpubl. report in files of Wyo. Game & Fish Comm., Cheyenne.

————— 1942. Sage grouse coccidia not transmissible to chickens. Poultry Sci. 21 (6).

————— 1947. Coccidiosis as a decimating factor among sage grouse. Final report Wyo. F. A. proj. 33-R. Wyo. Game & Fish Comm. : 1-12.

————— 1949. A survey to determine the importance of coccidiosis as a decimating factor among sage grouse in the Shirley Basin area. Quarterly progress report Wyo. F. A. proj. 33-R-2. Wyo. Game & Fish Comm. : 1-8.

Honess, Ralph F., and Warren J. Allred. 1942. Sage grouse studies. Part I. Structure and function of the neck muscles in inflation and deflation of the esophagus in the sage grouse. Wyo. Game & Fish Dept. Bull. No. 2 : 1-12.

Horsfall, B. B. 1932. A morning with sage grouse. Nature Magazine 20 (5) : 205.

Hull, A. C., Jr., Clyde W. Doran, C. H. Wasser, and D. F. Hervey. 1950. Reseeding sagebrush lands of western Colorado. Colo. Agric. and Mech. Coll. Bull. 413-A : 1-23.

Huntington, D. W. 1897. The sage grouse. The Osprey 2 (2) : 17-18.

Johnson, Donald E. 1951. Biology of the elk calf, *Cervus canadensis nelsoni*. Jour. Wildl. Mgt. 15 (4) : 396-410.

Keller, Robert J., Harold R. Shepherd, and Robert N. Randall. 1941. Report of the sage grouse survey in Colorado. Colo. Game & Fish Comm. Publ. : 1-31.

Kimball, James W. 1949. The crowing count pheasant census. Jour. Wildl. Mgt. 13 (1) : 101-102.

Knight, S. H., and O. A. Beath. 1937. The occurrence of selenium and seleniferous vegetation in Wyoming. Univ. of Wyo. Agric. Expt. Sta. Bull. 221.

Lang, Robert. 1945. Density changes of native vegetation in relation to precipitation. Wyo. Agric. Expt. Sta. Bull. 272 : 1-31.

Lehmann, Valgene W. 1941. Attwater's prairie chicken, its life history and management. U. S. Fish & Wildl. Serv. N. Amer. Fauna 57 : 1-65.

Ligon, J. Stokely. 1946. Upland game bird restoration through trapping and transplanting. Univ. New Mex. Publications in Biology No. 2 : 1-77.

Lommasson, T. 1947. Transition of sagebrush, *Artemisia tridentata*, on the semiarid foothills of the upper Missouri River Valley following burning. Developments in Range Management No. 6, U. S. Forest Service, Missoula, Mont., : 1-2.

      1948. Succession in sagebrush. Jour. Range Mgt. 1 (1) : 19-21.

Martin, Alexander C., Herbert S. Zim, and Arnold L. Nelson. 1951. American wildlife and plants. McGraw-Hill Book Co., Inc., New York : 1-500.

Miles, Wayne H. 1946. Clearing and reseeding sagebrush lands. U. S. Soil Cons. Serv. Regional Bull. No. 100, Range Mgt. Series No. 10. Albuquerque : 1-16.

Moffitt, James. 1942. Apparatus for marking wild animals with colored dyes. Jour. Wildl. Mgt. 6 (4) : 312-318.

Northrup, King & Company. 1949. Westland pastures for the intermountain region. Salt Lake City.

Parker, R. R., Cornelius B. Phillip, and Gordon E. Davis. 1932. Tularemia : occurrence in the sage hen, *Centrocercus urophasianus*. Public Health Reports 47 (9) : 479-487.

Post, George. 1950a. Crop contents studies of several game birds found during the summer of 1949. Progress report Wyo. F. A. Proj. 28-R-3. Wyo. Game & Fish Comm. : 1-6.

      1950b. Observations of deleterious effects of Toxaphene and Chlordane insecticides on game birds. Progress report Wyo. F. A. proj. 28-R-4. Wyo. Game & Fish Comm. : 1-10.

      1951a. The effects of Aldrin insecticide on birds. Final report Wyo. F. A. proj. 28-R-5. Wyo. Game & Fish Comm. : 1-13.

      1951b. Effects of Toxaphene and Chlordane on certain game birds. Jour. Wild. Mgt. 15 (4) : 381-386.

Price, Raymond. 1948. Conservation problems and practices. Grass—The Yearbook of Agriculture. U. S. D. A. : 569-574.

Price, Raymond, Kenneth W. Parker, and A. C. Hull, Jr. 1948. Range practices and problems. Grass—The Yearbook of Agriculture. U. S. D. A. : 557-560.

Rasmussen, D. I., and Lynn A. Griner. 1938. Life history and management studies of the sage grouse in Utah, with special reference to nesting and feeding habits. Trans. N. Amer. Wildl. Conf. 3 : 852-864.

Savage, D. A. 1937. Drought survival of native grass species in the central and southern Great Plains. U. S. D. A. Tech. Bull. 549.

Schwartz, Charles W. 1945. The ecology of the prairie chicken in Missouri. The Univ. of Mo. Studies 20 (1) : 1-99.

Scott, John W. 1940. The role of coccidia as parasites of wildlife. Jour. Colo.-Wyo. Acad. Sci. 2 (6) .

___________ 1942. Mating behavior of the sage grouse. Auk 59 : 477-498.

Simon, Felix. 1939. *Eimeria centrocerci* n. sp. du *Centrocercus urophasianus,* (coq de bruyene) . Annales de Parasitology, Humaine et Comparee, Tome XVII (2) .

___________ 1940. The parasites of the sage grouse. Univ. of Wyo. Publications 7 (5) : 77-100.

Simon, James R. 1940. Mating performance of the sage grouse. Auk 57 : 467-471.

Smith, Arthur D. 1950. Sagebrush as a winter feed for deer. Jour. Wildl. Mgt. 14 (3) : 285-289.

Sowls, Lyle K. 1948. The Franklin ground squirrel, *Citellus franklinii* (Sabine) , and its relationship to nesting ducks. Jour. Mamm. 29 (2) : 113-137.

Swainson, William, and John Richardson. 1831. Fauna Boreali-Americana. Part 2. The birds. J. Murray, Publishers, London : 358-361.

Vass, A. F., and Robert Lang. 1938. Vegetative composition, density, grazing capacity and grazing land values in the Red Desert area. Wyo. Agric. Expt. Sta. Bull. No. 229 : 1-72.

Visher, S. S. 1913. Notes on the sage hen. The Wilson Bulletin 25 (2) : 90-91.

Ward, Justus C., Malcolm Martin, and Warren J. Allred. 1942. The susceptibility of sage grouse to strychnine. Jour. Wildl. Mgt. 6 (1) : 55-57.

Weaver, John E., and Frederic E. Clements. 1938. Plant ecology. McGraw-Hill Book Co., Inc., New York : 1-601.

Wetmore, Alexander. 1951. Secretary, Smithsonian Institution. Personal letter, May 8, 1951.

Wyoming Agricultural Experiment Station. 1938. Vegetative composition, density, grazing capacity, and grazing land values in the Red Desert area. Wyo. Agric. Expt. Sta. Bull. No. 229 : 1-72.

INDEX